Protected sta
Species protec
included on a F

Five petals

Marsh-marigold

Caltha palustris
Buttercup family (Ranunculaceae)
MARCH–JUNE 15–50CM ▼ ☙

Note Flowers shiny golden-yellow.
Description Flower stalks curve upwards. Leaves heart- or kidney-shaped, to 15cm diameter.
Distribution Throughout the Alps from lowlands to 2400m. Wet or peaty soils. Streamsides and flushes. Common.

Hybrid Buttercup

Ranunculus hybridus
Buttercup family (Ranunculaceae)
JULY–AUG 5–15CM ▼ ☙

Note Basal leaves toothed like a cockscomb.
Description Flowers small, to 1.5cm diameter. Basal leaves usually 1–2, fleshy-leathery, often frosted.
Distribution Northern and southern ranges of eastern Alps to 2500m, calcicole. Snow-patches, scree and rock crevices. Scattered.

Lesser Spearwort

Ranunculus flammula
Buttercup family (Ranunculaceae)
JULY–OCT 15–50CM ☙

Note All leaves narrowly-lanceolate and entire.
Description Growth form varies, branching above. Flowers pale yellow, 0.5–2cm diameter. Lower leaves long-stalked.
Distribution Wet soils subject to occasional drying. Meadows, pastures and mires. Common below 1500m, above this rarer.

Photo shows typical appearance of plant in flower.

Alpine regions in which the species grows. Scattered distribution indicated by dots; even distribution by shading (see map at the end of the book).

Note shows key characters to aid identification.

Description Further features useful for identification.

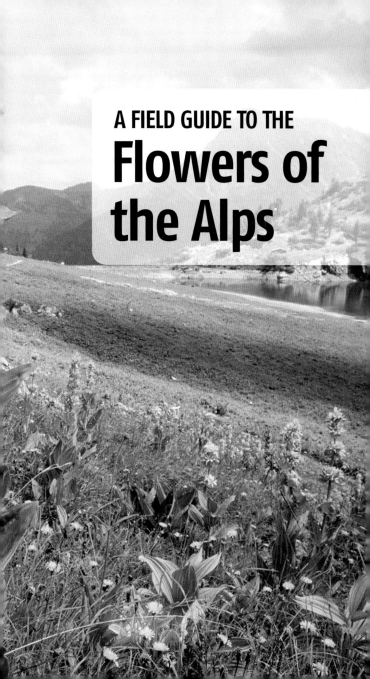

A FIELD GUIDE TO THE
Flowers of the Alps

Contents

An introduction to alpine flowers 6

Alpine flowers by flower colour 10–181

Red flowers 10

White flowers 52

Blue flowers 100

Yellow flowers 132

Green/brown flowers 176

Index 182
Botanical terms illustrated 190

The Alps – map of the regions covered
(on inside back cover)

An introduction to alpine flowers

Tall-herb communities develop particularly at sites where livestock rest or gather. These plants mostly flower in summer (here we see Monk's-hood and Alpine Ragwort)

The Alps are richer in colourful flowers than almost any other European region and the alpine flora is particularly fascinating because of the extreme climatic conditions found there. Above the tree line in particular, alpine flowers are especially showy in order to attract pollination by the relatively few insects that live at high altitudes. Alpine flowers therefore create a uniquely colourful display during the short spring and summer. This book includes more than 500 alpine flowers. Hill-walkers, tourists, climbers and others who wish to be able to recognise both common and more local species will find them here, and each species described can be identified without specialist botanical knowledge. The arrangement according to flower colour and shape makes it easier to identify the species and to compare the illustrations with the descriptions. The distinctive characters of each flower are described, allowing precise identification in comparison with similar species.

This is further aided by glancing at the symbols and the map of Alpine regions, the latter showing the rough divisions of the Alps according to plant geography and geology: some species for example only occur in the western or eastern Alps, and some closely related species are restricted to either calcareous or acidic soils. The border between the western and eastern Alps follows the line: Lake Como – Rhine valley – Lake Constance (Bodensee). The eastern Alps may be divided into three regions: the northern and southern calcareous Alps, and between them the central Alps whose rocks are mainly acidic. This central zone continues to the southwest, where it contains the highest of the alpine peaks. To the north and west lie the northern calcareous Alps, whose soils are mainly lime-rich. Solid shading indicates the species is common, while a dot indicates a scattered distribution.

In areas that have been grazed for long periods there are often stands of tall herbs that thrive on the dung-enriched soil. Above this we find a zone with dwarf shrubs – dominated by Mountain Pine (*Pinus mugo*) or Green Alder (*Alnus viridis*), grading higher still into low heath and alpine grassland, with occasional damp spring-water flushes or snow-patch vegetation around areas of long snow cover. Between 2800 and 3000m we reach the end of closed vegetation. Here low-growing cushion plants cling to the loose stones and scree or grow in rock crevices or gullies, to altitudes

Glacier Buttercup is adapted to grow in the extreme conditions of high altitude.

Purple Saxifrage holds the altitudinal record for a European flowering plant.

From alpine foothills to rocky heights

Climbing from the valleys to the alpine peaks, one passes through climatically determined altitudinal zones, each with its own characteristic vegetation, though often with indistinct boundaries. The mainly cultivated lowlands and valleys give way to montane forest, at first broadleaved deciduous, and eventually coniferous. The limit of closed woodland lies between 1800 and 2200m above sea level, though this has often been lowered through clearance for pasture.

of over 4000m. Beyond this one finds only mosses and lichens.

Many alpine flowers are legally protected and may not be picked, and collecting plants from nature reserves and national parks is also forbidden. Some species have become rare through human impact and have Red List status. These and all other species and their habitats are worthy of special protection in order to preserve the colourful world of the alpine flora.

Alpine Flowers

Four petals

Alpine Meadow-rue

Thalictrum alpinum
Buttercup family (Ranunculaceae)
JUL–AUG. 5–15CM ▼ (☠)

Note Yellow stamens extend beyond corolla.
Description Perianth segments 4–5, reddish-brown. Flowers pendent. Stamens with violet anthers, leaves 2-ternate, leaflets rounded and notched.
Distribution West, central and southern Alps to 2800m. Damp, stony sites; also boggy ground. Rare.

Greater Meadow-rue

Thalictrum aquilegifolium
Buttercup family (Ranunculaceae)
MAY–JUNE 50–120CM ▼

Note Flowers with fluffy-looking groups of stamens.
Description Petals tiny, falling early. Leaves resembling those of aquilegia, 2-3 ternate, notched.
Distribution Between 900 and 2500m. Mainly in the calcareous regions; common on moist soils.

Mountain Sorrel

Oxyria digyna
Dock family (Polygonaceae)
JULY–AUG. 5–15CM

Note Leaves kidney-shaped and rounded.
Description Flowers inconspicuous, reddish-green, in loose branching spikes. Leaves long-stalked.
Distribution Acid, stony soils with long snow cover, scree, and stony grassland. Common.

Four petals

Pyrenean Whitlowgrass

Petrocallis pyrenaica
Crucifer family (Brassicaceae)
JUNE–JULY 2–8CM ▼

Note Leaves three-lobed.
Description Cushion plant. Flowers to 1cm diameter, pleasantly-scented. Stem leafless. Leaves 3–5 lobed with erect hairs.
Distribution Mainly in the northern and southern calcareous Alps, on dry rocky soils between 1700 and 3500m. Rare.

Round-leaved Penny-cress

Thlaspi cepaeifolium
Crucifer family (Brassicaceae)
JUNE–JULY 5–15CM

Note Leaves rounded and entire.
Description Forms loose mats. Flowers pink or violet, 8–18mm diameter. Stem with leaves, lower leaves form rosette.
Distribution Across the Alps between 1500 and 3300m, loose rocks and scree. Common in calcareous Alps, otherwise rare.

Spring Heath

Erica carnea
Heath family (Ericaceae)
MARCH–JUNE 10–30CM ▼

Note Narrow bell-shaped flowers droop to one side.
Description Calyx shorter than the 4-lipped corolla. Violet anthers protruding. Leaves needle-shaped, in whorls of 3–4, margins rolled inwards.
Distribution In light coniferous woods and dwarf-shrub heath to 2700m. Common.

Four petals

Great Burnet

Sanguisorba officinalis
Rose family (Rosaceae)
JULY–SEPT. 30–90CM

Note Cylindrical-oval inflorescence with 20–40 flowers.
Description Flowers tiny with just 4 sepals. Leaves pinnate, with 7–15 toothed leaflets.
Distribution Between 400 and 2400m, on damp or occasionally damp meadows and fens. Common.

Garland Flower

Daphne cneorum
Daphne family (Thymeleaceae)
MAY–JULY 10–30CM ▼ ☠

Note Leaves regularly arranged on stem.
Description Dwarf-shrub with branching, hairy stems. Flowers in clusters of 6–10, smelling of aniseed. Corolla tube uniformly coloured, hairy.
Distribution On calcareous soil to 2000m. Dwarf-shrub heath, open woodland and scrub, rocky sites. Rare.

Striped Daphne

Daphne striata
Daphne family (Thymeleaceae)
MAY–AUG. 5–15CM ▼ ☠

Note Leaves towards tip of shoot.
Description Dwarf shrub with hairless stems. Flowers in tight clusters of 8–15, lilac-scented. Flower stems striped and hairless.
Distribution Mainly in the eastern Alps on calcareous soil to 2500m, on nutrient-poor, stony sites. Rare.

Four petals

Chickweed Willowherb

Epilobium alsinifolium
Willowherb family (Onagraceae)
JUNE–AUG. 5–25CM

Note Leaves are shiny and rather thick.
Description Petals 4, slightly notched with red veins. Leaves opposite, 2–4cm long, shallowly toothed.
Distribution Between 1200 and 2900m on wet soils: spring-fed flushes, banks, and tall-herb communities. Scattered to common.

Rosebay Willowherb

Epilobium angustifolium
Willowherb family (Onagraceae)
JUNE–AUG. 50–150CM

Note Flower spikes tall, many-flowered.
Description Flowers to 2.5cm diameter. Leaves alternate, linear-lanceolate, 5–20cm long, blue-green below. Edible when young.
Distribution Between 400 and 2500m. Clearings, scrub, scree. Common.

Fleischer's Willowherb

Epilobium fleischeri
Willowherb family (Onagraceae)
JULY–AUG. 10–30CM ▼

Note Flowers large, 2–4cm diameter.
Description Flowers in terminal clusters of 5–10. Leaves alternate, closely packed, only 1–4mm broad.
Distribution Mainly in western Alps, east of Tyrol. To 2500m. River gravel and stony soils. Scattered.

Four petals

Alpine Willowherb

Epilobium alpestre
Willowherb family (Onagraceae)
JUNE–AUG. 30–80CM

Note Leaves in whorls of 3 or 4.
Description Stem has 2–4 rows of hairs. Leaves broadly-lanceolate, shiny above.
Distribution Between 1300 and 2400m. Damp, rich soils. Tall-herb communities, Alder scrub. Common.

Field Gentian

Gentianella campestris
Gentian family (Gentianaceae)
JULY–SEPT. 5–20CM ▼

Note Calyx free almost to base.
Description Flowers 1–3cm long, pink-violet, with central tuft of hairs, in leaf axils and terminal. Leaves often drop before flowers appear.
Distribution Acid and nutrient-poor soils, 1000–2000m. Meadows and pastures. Scattered.

Sub-shrub Speedwell

Veronica fruticulosa
Figwort family (Scrophulariaceae)
JUNE–AUG. 10–20CM

Note Pink petals have red veins.
Description Develops woody runners. Stems with dense, opposite leaves. Leaves narrow oval or linear.
Distribution West and southern Alps on limestone, to 2400m. Absent from north-eastern Alps. Rocky clefts and rubble. Scattered.

Five petals

Alpine Pink

Dianthus alpinus
Carnation family (Caryophyllaceae)
JUNE–AUG. 3–15CM ▼

Note Dark ring at centre of flower.
Description Flowers flecked white, 1–2cm diameter, mostly single-stemmed. Leaves linear-lanceolate, 5mm broad.
Distribution In north-eastern Alps to 2400m, on limestone. Stony, loose grassland, and scrub. Rare.

Sweet William

Dianthus barbatus
Carnation family (Caryophyllaceae)
JUNE–SEPT. 20–70CM ▼

Note Flowers in dense terminal clusters.
Description Clusters of up to 30 flowers, surrounded by long, pointed bracts. Leaves to 2cm broad.
Distribution South-eastern Alps between 1000 and 2500m. Dry calcareous and nutrient-rich soils. Rare, sometimes naturalized (common garden plant).

Glacier Pink

Dianthus glacialis
Carnation family (Caryophyllaceae)
JULY–AUG. 2–8CM ▼

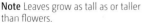

Note Leaves grow as tall as or taller than flowers.
Description Flowers purple-red, 1–2cm diameter, single-stemmed. Leaves narrow, to 5cm long.
Distribution Only in eastern central Alps, to 2900m. Rocky, exposed sites. Rare.

Five petals

Large Pink

Dianthus superbus
Carnation family (Caryophyllaceae)
JUNE–SEPT. 20–70CM ▼

Note Petals pink and deeply cut.
Description Flowers large, 3–6cm diameter, sweetly scented. Leaves narrow, to 10mm broad.
Distribution Across entire Alps, in several subspecies, between 300 and 2400m. Grassland, scrub and woodland edge. Scattered.

Wood Pink

Dianthus sylvestris
Carnation family (Caryophyllaceae)
JUNE–SEPT. 10–50CM ▼

Note Forms small cushions.
Description Petals pink to reddish-violet, unmarked, toothed. Leaves to 4cm long, 2mm broad, grooved.
Distribution Entire Alps to 2800m. Nutrient-poor, dry, rocky soils, stony grassland, rocky debris. Scattered.

Rock Soapwort

Saponaria ocymoides
Carnation family (Caryophyllaceae)
MAY–SEPT. 10–35CM ▼

Note Corolla with glandular hairs.
Description Spreading growth. Flowers clustered on branching stems. Leaves narrow, softly hairy at base.
Distribution Entire Alps to 2300m, but commonest in western Alps on calcareous soils. Scattered to common.

Five petals

Dwarf Soapwort

Saponaria pumila
Carnation family (Caryophyllaceae)
AUG–SEPT. 3–8CM ▼

Note Petals widely separated.
Description Cushion forming. Flowers to 2.5cm broad, single-stemmed. Leaves narrow and somewhat fleshy.
Distribution In central eastern Alps to 2600m. Acid, stony grassland and dwarf-shrub heath. Scattered.

Moss Campion

Silene acaulis
Carnation family (Caryophyllaceae)
JUNE–SEPT. 1–5CM ▼

Note Forms thick cushions.
Description Flowers to 2cm across, short-stemmed. Leaves to 12mm long, leathery and fringed with hairs.
Distribution Mainly in the outer Alpine ranges between 1700 and 2900m. Base-rich soils, poor grassland, rocky and stony sites. Scattered to common.

Red Campion

Silene dioica
Carnation family (Caryophyllaceae)
MAY–SEPT. 30–90CM

Note Dioecious (male and female flowers on separate plants).
Description Flowers to 3cm across, calyx narrow with 10 veins (male) or inflated and with 20 veins (female). Leaves broadly-lanceolate and hairy.
Distribution To 2300m. Damp, rich soils, tall-herb communities, scrub. Common.

Five petals

Alpine Catchfly

Silene suecica
Carnation family (Caryophyllaceae)
JULY–AUG. 5–20CM ▼

Note Flowers in tight terminal clusters.
Description Flowers purple-red, about 1cm across. Stem unbranched, not sticky. Leaves narrowly-lanceolate.
Distribution In central Alps between 2300 and 3000m. Dry, acid, exposed soils, ridges, stony grassland. Rare.

Common Bistort

Bistorta officinalis
Dock family (Polygonaceae)
JUNE–JULY 30–80CM

Note Small flowers in dense terminal spikes.
Description Leaves bluish-green and softly hairy below. Medicinal and food plant.
Distribution To 2500m. Wet nutrient-rich soils, damp meadows and pasture. Common.

Alpine Thrift

Armeria alpina
Thrift family (Plumbaginaceae)
JULY–AUG. 5–20CM ▼

Note Leaves grass-like.
Description Grows in dense cushions. Dense, rounded flowerheads with papery bracts.
Distribution Mainly in central and southern ranges to 3000m, on grassland or in rocky scree. Rare.

Five petals

Trailing Azalea

Loiseleuria procumbens
Heath family (Ericaceae)
JUNE–JULY 15–30CM ▼

Note Small shrub, low-growing.
Description Terminal clusters of 2–5 flowers. Leaves long and needle-like, with inrolled edges, shiny above.
Distribution 1500–3000m on acid soils. Exposed ridges, scree. Common in central Alps, otherwise rare.

Alpenrose

Rhododendron ferrugineum
Heath family (Ericaceae)
JUNE–JULY 20–100CM ▼

Note Leaves rusty red and mealy beneath.
Description Flowers bright red, about 1.5cm long, bell-shaped, 5 petals with pointed tips, fused into tube. Leaves about 4cm long, oval, leathery and hairless, dark green and shiny above.
Distribution Acid soils, mainly in the inner mountains, to 2500m. Woods and dwarf-shrub heath. Common.

Hairy Alpenrose

Rhododendron hirsutum
Heath family (Ericaceae)
JUNE–JULY 20–80CM ▼

Note Leaves have hairy margins.
Description Flowers pale red, bell-shaped, about 1.5cm long. Leaves to 3cm long, oval, leathery, and shiny above.
Distribution Damp calcareous soils, 1500–2500m. Woods and dwarf-shrub heath. Common in calcareous Alps. Elsewhere rare.

Five petals

Dwarf Alpenrose

Rhodothamnus chamaecistus
Heath family (Ericaceae)
MAY–JULY 10–40CM ▼

Note Flowers open out flat.
Description Flowers 1–3, terminal, long-stemmed. Leaves oval-lanceolate, about 1cm long, hairy, with leathery margins.
Distribution In eastern Alps on calcareous soils, between 1500 and 2200m, west as far as the Allgäu. Rocky cliffs, scree. Widespread.

Flesh-pink Rock-jasmine

Androsace adfinis ssp. *puberula*
Primrose family (Primulaceae)
JUNE–AUG. 3–12CM ▼

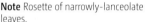

Note Rosette of narrowly-lanceolate leaves.
Description Flowers 1–8, on tall stems, pink, darker inside with yellow centre. Other subspecies have whiter flowers.
Distribution South-western Alps as far as Wallis, between 1800 and 3000m, on acid, stony soils. Rare.

Alpine Rock-jasmine

Androsace alpina
Primrose family (Primulaceae)
JULY–AUG. 2–5CM ▼

Note Leaf rosettes form mat-like cushions.
Description Flowers 5–9mm diameter, individual, short-stemmed. Leaves covered with whitish hairs.
Distribution Mainly on inner ranges between 2000 and 3400m, rarely to over 4000m. Damp acid soils. Scattered to rare.

Five petals

Alpine Bells

Cortusa matthioli
Primrose family (Primulaceae)
MAY–AUG. 20–50CM ▼ (☠)

Note Flowers droop to one side.
Description Flowers 5–10, in umbels, on long hairy stems. Leaves large, to 12cm across, with long stalks.
Distribution Damp, acid soils between 1000 and 2000m, often in shade. Scrub, open woodland, tall-herb communities. Very rare.

Alpine Cyclamen

Cyclamen purpurascens
Primrose family (Primulaceae)
JUNE–SEPT. 5–20CM ▼ ☠

Note Petal lobes reflexed.
Description Flowers to 2.5cm long on leafless stems. Leaves heart-shaped, stalked, with pale markings. Traditional medicinal plant.
Distribution Southern and south-eastern Alps to 2000m. Stony, calcareous soils. Pine woods, Mountain Pine scrub. Rare.

Bird's-eye Primrose

Primula farinosa
Primrose family (Primulaceae)
MAY–JULY 5–25CM ▼

Note Calyx almost as long as corolla tube.
Description Flowers in many-flowered umbels. Flower stems much longer than leaves. Leaves mealy white beneath.
Distribution On damp, nutrient-rich, mainly calcareous soils to 2700m. Mires, spring-fed flushes, grassland. Scattered.

Five petals

Haller's Primrose

Primula halleri
Primrose family (Primulaceae)
JUNE–JULY 10–30CM ▼

Note At 2–3cm, the corolla tube is longer than the calyx.
Description Umbels of several flowers. Stalk much longer than the leaves. Leaves mealy white beneath.
Distribution Mainly southern and south-eastern Alps to 2900m. Stony, calcareous soils. Rare.

Hairy Primrose

Primula hirsuta
Primrose family (Primulaceae)
MAY–JULY 3–10CM ▼

Note Leaves with short, sticky glandular hairs.
Description Flowers 1–5 on the stalk, the latter usually shorter than the grooved fleshy leaves.
Distribution Inner Alpine ranges on acid soils to over 3000m. Rocky crevices, rubble and stony grassland. Scattered to common.

Entire-leaved Primrose

Primula integrifolia
Primrose family (Primulaceae)
JUNE–JULY 2–5CM ▼

Note Small, entire leaves are almost stalkless.
Description Flowers to 2.5cm diameter, 1–6 per stalk. Leaves with softly hairy margins.
Distribution Central ranges of western Alps, east to a line from Tonale–Arlberg. Often in snow hollows. Scattered.

Five petals

Least Primrose

Primula minima
Primrose family (Primulaceae)
JUNE–JULY 1–4CM ▼

Note Flowers twice as large as leaves.
Description Flowers to 3cm diameter, petals cut to half way. Leaves shiny, with 3–9 teeth around margin.
Distribution Eastern Alps on acid soils to 3000m. Snow hollows, rock clefts and rubble. Scattered.

Piedmont Primrose

Primula pedemontana
Primrose family (Primulaceae)
MAY–JULY 5–20CM ▼

Note Leaf margins have small red glands.
Description Flowers in umbels of 2–8, on stalks twice as long as leaves. Leaves 2–6cm long, entire or wavy, not shiny.
Distribution Western Alps to 3000m, on acid humus-rich soils. Rock clefts and rubble. Very rare.

Splendid Primrose

Primula spectabilis
Primrose family (Primulaceae)
MAY–JULY 5–20CM ▼

Note Leaves shiny, with obvious white edge.
Description Flowers 2–3cm diameter, in umbels of 2–7. Calyx with pointed teeth. Leaves 3–9cm long.
Distribution Southern Alps between 700 and 2200m on calcareous soils. Stony grassland, rock crevices and rubble. Rare.

Five petals

Love-restoring Stonecrop

Sedum anacampseros
Stonecrop family (Crassulaceae)
JULY–AUG. 10–25CM

Note Flowers in dense, rounded clusters.
Description Flowers 5–7mm diameter on leafy stems. Leaves stalkless, flat, entire, somewhat fleshy.
Distribution From Maritime Alps to Wallis (Valais) and in Trentino, between 1400 and 2500m. Rocky, acid soils. Scattered.

Dark Stonecrop

Sedum atratum
Stonecrop family (Crassulaceae)
JUNE–AUG. 3–8CM ▼

Note Flowers on all shoots.
Description Flowers mainly red, or with red stripe on yellowish petals. Leaves red, hairless, almost round in cross-section.
Distribution In calcareous Alps to 3100m. Rocks, rubble and scree. Common in calcareous areas, otherwise rare.

Two-flowered Saxifrage

Saxifraga biflora
Saxifrage family (Saxifragaceae)
JULY–AUG. 2–10CM ▼

Note Narrow petals.
Description Flowers 2–9 on short leafy stems. Leaves opposite, fleshy, to 5mm long.
Distribution Damp, calcareous soils between 2000 and over 4000m. Moraines, scree. Scattered to rare.

Five petals

Orange Saxifrage

Saxifraga mutata
Saxifrage family (Saxifragaceae)
JULY–AUG. 10–40CM ▼

Note Large leaf rosettes – 5–15cm across.
Description Flowers yellow to deep orange, with red centre, in panicle.
Distribution Calcareous Alps to 2200m. Damp calcareous soils. Scattered; rare in central Alps.

Purple Saxifrage

Saxifraga oppositifolia
Saxifrage family (Saxifragaceae)
APRIL–JULY 2–5CM ▼

Note Flowers 1–2cm diameter – considerably larger than leaves.
Description Forms loose cushions. Flowers solitary. Leaves opposite, often with chalky encrustation.
Distribution Above 800m – holds European altitude record at 4505m. Damp, stony soil, stony grassland, scree, rocks. Common.

Dolomite Cinquefoil

Potentilla nitida
Rose family (Rosaceae)
JUNE–SEPT. 2–8CM ▼

Note Flowers usually pinkish-red.
Description Forms low cushions. Petals with slightly wavy edges. Leaves usually three-lobed.
Distribution Almost entirely restricted to the southern Alps between 1200 and 3100m, on dry calcareous rocky soils. Rocks and rubble. Rare.

Five petals

Dusky Crane's-bill

Geranium phaeum
Crane's-bill family (Geraniaceae)
JUNE–JULY 30–60CM ▼

Note Flowers spreading, brownish-violet.
Description Flowers to 2.5cm diameter. Leaves divided to just over halfway, with up to 7 deeply-toothed lobes.
Distribution Mainly in central ranges, between 1000 and 2000m. Damp nutrient-rich soils. Meadows, tall-herb communities. Scattered to rare.

Bloody Crane's-bill

Geranium sanguineum
Crane's-bill family (Geraniaceae)
MAY–JULY 20–60CM ▼

Note Flowers comparatively large, to 4cm diameter.
Description Flowers deep red, solitary, stalked. Leaves deeply and narrowly lobed
Distribution Between 400 and 1800m. Warm, dry soils, normally calcareous. Open woodland and scrub. Scattered to rare.

Wood Crane's-bill

Geranium sylvaticum
Crane's-bill family (Geraniaceae)
JUNE–AUG. 30–70CM

Note Sepals with long awns.
Description Flowers with whitish centres, 1.5–2.5cm diameter. Leaves deeply divided into 6 lobes.
Distribution Between 700 and 2500m. Rich, damp soils. Mountain meadows, tall-herb communities, open woodland and scrub. Common.

Five petals

Alpine Lovage

Ligusticum mutellina
Umbellifer family (Apiaceae)
JUNE–AUG. 10–80CM

Note Brownish tuft of fibres at base of stem.
Description Umbel bracts fall early. Leaves 2–3-pinnate, with aromatic scent. Used to flavour spirits and as medicinal plant.
Distribution Rich soils to 2600m. Meadows, pasture and tall-herb communities. Common.

Unbranched Lovage

Ligusticum mutellinoides
Umbellifer family (Apiaceae)
JUNE–AUG. 3–15CM

Note Umbel bracts large and retained.
Description Flowers pink or white. Stem leafless or with 1 leaf. Leaves 2–3-pinnate, oval in outline.
Distribution Stony, acid soils between 2000 and 3000m. Scree, stony grassland, and ridges.

Great Burnet-saxifrage

Pimpinella major
Umbellifer family (Apiaceae)
JUNE–SEPT. 30–80CM ▼

Note Leaf lobes almost triangular in outline.
Description Flowers pinkish or white. Stems strongly ridged. Leaves simply pinnate, with 3–9, pointed, toothed lobes. Ancient medicinal plant.
Distribution Between 600 and 2200m. Mountain meadows and tall-herb communities. Common.

Five petals

Devil's Claw

Physoplexis comosa
Bellflower family (Campanulaceae)
JUNE–AUG. 5–15CM ▼

Note Flowers have violet tips, pointed and claw-like.
Description Flowerheads of 10–30 flowers. Leaves kidney-shaped or oval, toothed.
Distribution Only in southern calcareous Alps to 2000m. Rock clefts and debris. Very rare.

Mountain Valerian

Valeriana montana
Valerian family (Valerianaceae)
MAY–AUG. 10–50CM

Note Flowers pale pink, in umbel-like clusters.
Description Inflorescence many-flowered. Flowers to 6mm diameter. Stems with 3–8 pairs of leaves. Leaves shiny, entire and shallowly toothed.
Distribution Stony calcareous soils between 1000 and 2600m. Common in calcareous Alps, otherwise rare.

Dwarf Valerian

Valeriana supina
Valerian family (Valerianaceae)
JUNE–AUG. 3–12CM ▼

Note Flowerheads (fewer flowers than previous species), surrounded by bracts.
Description Flowers pale pink, 2–5mm long. Stems with 1–2 pairs of leaves. Leaves entire, somewhat fleshy.
Distribution Eastern Alps to 2700m, west to Graubünden. Damp, calcareous soils. Snow-patches, debris. Rare.

Many petals

Mountain Pasqueflower

Pulsatilla montana
Buttercup family (Ranunculaceae)
MARCH–MAY 10–30CM ▼ ☠

Note Leaf tips narrow.
Description Flowers dark violet to reddish, 5–6cm diameter. Leaves below bract whorl appear after flowers.
Distribution Southern and south-western Alps, 500–1600m, on very dry soils. Poor grassland. Rare.

Mountain Dock

Rumex alpestris
Dock family (Polygonaceae)
JUNE–AUG. 30–100CM

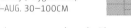

Note Leaves arrow-shaped with rounded tips.
Description Flowers reddish-green, in loose panicles. Leaves thin and soft, lower leaves long-stalked, almost triangular.
Distribution To 2200m. Rich, damp soils. Pasture, scrub. Scattered.

Snow Dock

Rumex nivalis
Dock family (Polygonaceae)
JULY–AUG. 5–25CM

Note Inflorescence red, cylindrical.
Description Flowers small, with reflexed petals. Leaves to 4cm long, rounded–oval to spear-shaped, somewhat fleshy.
Distribution Eastern Alps to 2500m. Damp, calcareous, stony soils. Rare.

Many petals

Monk's Rhubarb

Rumex alpinus
Dock family (Polygonaceae)
JUNE–AUG. 60–120CM

Note Leaves to 30cm, with wavy margins.
Description Many greenish-red flowers in dense panicles. Leaves long, oval, heart-shaped at base.
Distribution Throughout Alps to 2600m. Damp, over-fertilized sites. Mountain huts, cattle sheds. Common.

Sheep's Sorrel

Rumex acetosella
Dock family (Polygonaceae)
MAY–AUG. 10–30CM

Note Leaves narrow and pointed. Flowers small, red-green.
Description Female and male flowers on separate plants (plant is dioecious). Leaves sometimes not pointed.
Distribution To 2400m. Poor grassland, alongside paths. Common.

Common Peony

Paeonia officinalis
Peony family (Paeoniaceae)
MAY–JUNE 40–120CM ▼

Note Usually 8 large, purple-red solitary flowers.
Description Flowers to 15cm diameter. Leaves to 30cm long, 2-ternate.
Distribution Southern Alps, on dry, calcareous rocky soil to 1700m. Open woodland and scrub. Very rare in the wild; often cultivated.

Many petals

Heather

Calluna vulgaris
Heath family (Ericaceae)
JULY–OCT. 10–50CM

Note Flowers in racemes, mainly facing one side.
Description Petals 8. Leaves evergreen, opposite, 1–3mm long, c. 1mm wide, overlapping.
Distribution To 2900m on poor, acid, dry soils. Scattered to common.

Alpine Snowbell

Soldanella alpina
Primrose family (Primulaceae)
MAY–JULY 5–15CM ▼

Note Flowers deeply fringed – cut to about halfway.
Description Stems with 1–3 flowers. Flowers blue-violet–red-violet, funnel-shaped, nodding. Leaves rounded, with heart-shaped base, leathery.
Distribution Damp, rich soils, especially on limestone. To 2700m. Common.

Dwarf Snowbell

Soldanella pusilla
Primrose family (Primulaceae)
JUNE–AUG. 3–10CM ▼

Note Flowers fringed – cut to one quarter.
Description Flowers single-stemmed, pale violet, funnel-shaped, nodding. Leaves rounded-kidney-shaped, heart-shaped at base.
Distribution Poor, acid, stony soils. 2000–3000m. Snow-patches, damp grassland. Scattered.

Many petals

Cobweb Houseleek

Sempervivum arachnoideum
Stonecrop family (Crassulaceae)
MAY–SEPT. 5–15CM ▼

Note Tips of leaf rosettes covered with web-like hairs.
Description Flowers 8–12, 1–2cm diameter. Leaf rosette 1.5cm across.
Distribution Mainly central and southern Alps between 1000 and 2800m. Rocks, walls, rocky debris. Common to scattered.

Mountain Houseleek

Sempervivum montanum
Stonecrop family (Crassulaceae)
JULY–SEPT. 5–25CM ▼

Note Petals 12–16, red-violet.
Description Flowers 2–3cm diameter. Rosette leaves with red-brown tips and glandular hairs, lacking cilia at edges.
Distribution Stony acid soils between 1000 and 3000m. Rock clefts, rocky debris, and stony grassland. Scattered.

Alpine Houseleek

Sempervivum tectorum ssp. *alpinum*
Stonecrop family (Crassulaceae)
JULY–AUG. 10–40CM ▼

Note Rosette leaves have ciliated edges, otherwise glabrous.
Description Flowers 2–3cm diameter. Stem with glandular hairs. Leaf rosettes 3–8cm diameter, leaves with red-brown tips.
Distribution Between 800 and 2800m. Dry, stony soils. Rocks and poor grassland. Scattered..

Many petals

Hungarian Gentian

Gentiana pannonica
Gentian family (Gentianaceae)
JULY–AUG. 15–60CM ▼

Note Calyx teeth (5–8) arch outwards.
Description Flowers red-violet on outside, with dark spots, yellowish inside. Corolla cut almost to halfway.
Distribution Only in eastern Alps, to 2200m, on rich calcareous soils. Tall-herb communities, mires, Mountain Pine scrub. Scattered.

Purple Gentian

Gentiana purpurea
Gentian family (Gentianaceae)
JULY–AUG. 20–60CM ▼

Note Calyx papery, split to base on one side.
Description Flowers purple-red on outside, unspotted, yellowish inside. Corolla cut to a third.
Distribution Western Alps west to Allgäu, between 1600 and 2500m. Acid soils. Pasture, scrub, dwarf-shrub heath. Scattered.

Adenostyles

Adenostyles alliariae
Composite family (Asteraceae)
JUNE–SEPT. 50–150CM

Note Leaves with cobweb-like hairs beneath.
Description Flowerheads 3–6 flowered. Leaves to 50cm, heart-shaped, irregularly toothed.
Distribution Between 1000 and 2200m. on damp rich soils. Tall-herb communities, scrub and woods. Common.

Many petals

Alpine Adenostyles

Adenostyles alpina
Composite family (Asteraceae)
JUNE–AUG. 30–80CM

Note Undersides of leaves only hairy along veins.
Description Flowerheads mostly of 2–3 flowers. Leaves heart-shaped, regularly-toothed.
Distribution Mostly in the outer calcareous ranges, between 1000 and 2200m. Stony habitats and tall-herb communities, scrub and woods. Scattered.

Woolly Adenostyles

Adenostyles leucophylla
Composite family (Asteraceae)
JULY–AUG. 10–40CM ▼

Note Undersides of leaves covered with woolly hairs.
Description Flowerheads of 12–24 flowers. Stems softly hairy. Leaves kidney- to heart-shaped.
Distribution Western Alps, east to western Tyrol, between 2000 and 3000m. Acid, stony soils. Rockfalls, scree. Scattered.

Alpine Butterbur

Petasites paradoxus
Composite family (Asteraceae)
APRIL–JUNE 10–30CM

Note Leaves develop at flowering time. Leaves white-woolly beneath.
Description Flowerheads in racemes, to 8mm diameter. Stalks with red scales. Leaves triangular to heart-shaped.
Distribution Stony, calcareous soils to 2000m. Debris, river gravel. Common in calcareous Alps, otherwise rare.

Many petals

Mountain Everlasting

Antennaria dioica
Composite family (Asteraceae)
JUNE–JULY 5–25CM ▼

Note Flowerheads surrounded by dry, papery bracts.
Description Forms mats and produces slender stolons. Flowers red, pink, and white adjacent in flowerhead. Leaves oval to narrowly-lanceolate, cottony-white beneath. Ancient medicinal plant.
Distribution Dry, acid or calcareous soils. Heath, pasture. Common.

Purple Colt's-foot

Homogyne alpina
Composite family (Asteraceae)
MAY–AUG. 10–40CM

Note Kidney-shaped leaves are shiny above, often purplish beneath.
Description Flowerheads solitary at tip of stem, to 2.5cm diameter. Stem with woolly hairs.
Distribution Acid soils to 3000m. Open woodland, snow-patches, dwarf-shrub heath. Scattered.

Alpine Thistle

Carduus defloratus
Composite family (Asteraceae)
JUNE–SEPT. 15–70CM

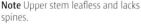

Note Upper stem leafless and lacks spines.
Description Flowerheads 2.5–4.5cm long, solitary, not (or only slightly) nodding. Leaves glabrous, toothed, edged with spines.
Distribution Dry, often calcareous soils, to 3000m. Pastures, open woods, heath. Scattered to common.

Many petals

Great Marsh Thistle

Carduus personata
Composite family (Asteraceae)
JUNE–AUG. 50–150CM ▼

Note Stems spiny.
Description Flowerheads 1.5–2.5cm long, in tight erect groups at end of stems. Leaves softly hairy beneath, lower leaves pinnately-lobed, upper leaves unlobed.
Distribution Damp nutrient-rich, mainly calcareous soils, 800–2500m. Tall-herb communities, including those on soils enriched by livestock dung, meadows. Scattered.

Alpine Greater Knapweed

Centauria scabiosa ssp. *alpestris*
Composite family (Asteraceae)
JULY–AUG. 30–70CM ▼

Note Flowerhead bracts black, triangular and fringed.
Description Flowerheads 1–3 and terminal. Leaves 1–2-pinnate, leathery.
Distribution Dry, calcareous soils. Meadows and scrub between 1200 and 2500m. Scattered, common in places.

Singleflower Knapweed

Centaurea uniflora
Composite family (Asteraceae)
JULY–SEPT. 5–30CM

Note Plant densely covered in soft white hairs.
Description Flowerheads solitary. Post-flowering heads have wig-like appearance. Leaves narrowly-lanceolate (mostly < 1cm across), entire or finely-toothed.
Distribution South-western Alps, only France and Italy. Mountain meadows to 2500m. Rare.

Many petals

Plumed Knapweed

Centaurea nervosa
Composite family (Asteraceae)
JULY–AUG. 10–40CM ▼

Note Plant grey-green and covered with rough hairs.
Description Flowerheads solitary and terminal. Post-flowering heads have wig-like appearance. Leaves lanceolate (mostly > 1cm across), toothed.
Distribution Mainly in central and southern Alps between 1000 and 2600m. Damp, acid soils. Scattered.

Wig Knapweed

Centaurea phrygia
Composite family (Asteraceae)
AUG–SEPT. 20–80CM ▼

Note Post-flowering heads have wig-like appearance.
Description Flowerheads 2–4 and terminal. Stems mostly branched. Leaves broadly-lanceolate, upper leaves clasping the stem, lower leaves stalked.
Distribution Damp, humus-rich soils. Meadows, scrub and woodland edges, to 2000m. Scattered.

Alpine Saw-wort

Saussurea rhapontica
Composite family (Asteraceae)
JULY–SEPT. 30–120CM ▼

Note Terminal flowerhead very large, 8–10cm diameter.
Description Stem thickened below flowerhead. Leaves to 50cm long, toothed, with white-cottony hairs beneath.
Distribution Damp, nutrient-rich and stony soils between 1400 and 2500m. Tall-herb communities and debris. Rare.

Many petals

Dwarf Thistle

Cirsium acaule
Composite family (Asteraceae)
JULY–SEPT. 5–15CM

Note Flowerheads stemless or very short-stemmed.
Description Flowers solitary in a large rosette. Leaves pinnately-lobed, spiny only at the edges.
Distribution Dry, calcareous soils. Meadows and pastures between 800 and 2300m. Generally common, though rare in the central Alps.

Woolly Thistle

Cirsium eriophorum
Composite family (Asteraceae)
JUNE–SEPT. 50–150CM ▼

Note Flowerheads to 7cm covered with cobweb-like hairs.
Description Bracts spiny. Leaves pinnate, with long, sharp spines, white-cottony beneath.
Distribution Nutrient-rich, dry soils to 2000m. Pasture, tracks, woodland edges. Scattered.

Melancholy Thistle

Cirsium helenioides
Composite family (Asteraceae)
AUG.–SEPT. 50–150CM ▼

Note Flowerhead bracts lack spines.
Description Stems and leaf undersides white-cottony. Leaves entire or with forward-pointing lobes. Plant not prickly.
Distribution Damp, nutrient-rich soils, 1200–2200m. Meadows, tall-herb communities, scrub. Common.

Many petals

Field Fleawort

Tephroseris integrifolia ssp. *capitata*
Composite family (Asteraceae)
JUNE–AUG. 15–40CM ▼

Note Flowers bright orange-red (in this subspecies).
Description Flowerheads densely packed 3–10, in terminal clusters. Leaves with grey-cottony hairs. Basal leaves short-stalked, stem leaves unstalked.
Distribution Only local in Alps, absent over large areas. Stony, calcareous soils. Rare.

Alpine Fleabane

Erigeron alpinus
Composite family (Asteraceae)
JULY–AUG. 5–20CM

Note Leaves hairy on both sides.
Description Flowerheads to 2.5cm diameter, solitary or in clusters of 2–5. Ray florets pink to red, tubular florets yellow. Leaves lanceolate.
Distribution Dry, poor, acid soils to 3000m. Pasture, grassland. Scattered to common.

Purple Lettuce

Prenanthes purpurea
Composite family (Asteraceae)
JULY–SEPT. 30–100CM

Note Flowerheads of only 3–5 ray florets.
Description Many flowerheads, in loose panicles. Leaves glabrous, at base clasping the stem. Milky sap.
Distribution Throughout Alps to 2000m. Nutrient-rich, weakly acid soils. Open woods and scrub, and tall-herb communities. Scattered.

Many petals

Golden Hawk's-beard

Crepis aurea
Composite family (Asteraceae)
JUNE–SEPT. 5–25CM

Note Flowerheads solitary and terminal.
Description Bracts and upper stem with black hairs. Leaves with rounded teeth or pinnate.
Distribution Damp, nutrient-rich, usually calcareous soils., 1200–2700m. Meadows and pastures. Scattered.

Orange Hawkweed

Hieracium aurantiacum
Composite family (Asteraceae)
JUNE–AUG. 20–50CM ▼

Note Whole plant covered with long hairs.
Description Flowerheads of 2–10, in umbel-like clusters, with only ray florets. Stem with 1–4 narrowly-lanceolate leaves. Milky sap.
Distribution Poor, acid soils, 1200–2500m. Meadows and pastures, stony grassland. Scattered.

Narcissus Onion

Allium narcissiflorum
Lily family (Liliaceae)
JULY–AUG. 10–40CM ▼

Note Flowers bell-shaped.
Description Flowers 3–8 in an umbel-like cluster. Stem rounded, upper stem compressed and 2-ridged. Leaves 3–5, flat.
Distribution South-western Alps on limestone between 1500 and 2000m. Scree and rocks. Scattered to rare.

Many petals

Chives

Allium schoenoprasum
Lily family (Liliaceae)
JUNE–AUG. 15–30CM

Note Leaves cylindrical and hollow.
Description Flowerheads globular, to 5cm diameter. Plant smells of garlic, or cultivated Chives.
Distribution Moist, stony soils to 2600m. Wet meadows, fens, and riverbanks. Scattered.

Fire Lily

Lilium bulbiferum
Lily family (Liliaceae)
MAY–JULY 20–100CM ▼

Note Large flowers, to 7cm, with dark spots on inside.
Description Stem has many narrowly-lanceolate leaves. Also cultivated.
Distribution Mainly south-western and southern Alps, to 2400m. Nutrient-rich, warm soils. Mountain meadows, edges of woodland, and scrub. Rare.

Dog's-tooth Lily

Erythronium dens-canis
Lily family (Liliaceae)
FEB.–APRIL 10–30CM ▼

Note Petals 6, bending backwards.
Description Flowers solitary on red stems. Leaves usually 2 long and lanceolate, with red-brown spots.
Distribution Only in southern Alps to 2000m on warm, dry, calcareous, stony soils. Open woods and scrub. Rare.

Many petals

Martagon Lily

Lilium martagon
Lily family (Liliaceae)
JUNE–AUG. 30–120CM ▼

Note Pink, purple-spotted tepals, bent back upwards and rolled inwards.
Description Flowers, in terminal racemes of 3–15, have dark, purplish spots. Leaves mostly in whorls.
Distribution Nutrient-rich, stony, usually calcareous soils to 2000m. Open woodland and scrub, tall-herb communities. Scattered.

Alpine Fritillary

Fritillaria tubiformis
Lily family (Liliaceae)
JULY–AUG. 10–30CM ▼

Note Purple-red flowers with chequered pattern.
Description Flowers solitary, terminal and drooping. Leaves grass-like, grey-green, 4–6 on stem, basal leaves lacking.
Distribution Only in south-western Alps, to 2100m, on calcareous and nutrient-rich grassland and meadows. Rare.

Alpine Saffron

Colchicum alpinum
Lily family (Liliaceae)
JULY–AUG. 7–15CM ☠

Note Lower part of flower forms a long tube.
Description Crocus-like flower with tepals 2–3cm long. Flowers in autumn before the leaves appear and fruits in spring. Very poisonous!
Distribution South-western Alps to 2000m. Meadows and pastures. Scattered to common.

Zygomorphic flowers

Mount Cenis Restharrow

Ononis cristata
Legume family (Fabaceae)
JUNE–SEPT. 5–20CM

Note Flowers bi-coloured – white keel and wings, and red and white striped standard.
Description Thornless, with glandular hairs. Leaves trifoliate with toothed lobes.
Distribution South-western Alps – only France and Italy. Amongst rocks and in open woodland to 2500m. Rare.

Alpine Clover

Trifolium alpinum
Legume family (Fabaceae)
JUNE–AUG. 5–20CM

Note Largest-flowered clover – flowers to 2.5cm long.
Description Flowers 3–15, in a loose cluster to 5cm across. Leaves glabrous, trifoliate, leaflets narrowly-lanceolate.
Distribution Mainly in central Alps on poor acid grassland, where common. Otherwise rare.

Noble Clover

Trifolium rubens
Legume family (Fabaceae)
JUNE–JULY 20–60CM ▼

Note Flowers tightly packed in spikes 3–7cm long.
Description Calyx teeth long, with ragged hairs. Leaves glabrous, trifoliate, leaflets narrowly-lanceolate.
Distribution Dry, sandy soils. Open woodland and scrub between 500 and 2000m, mainly on central and southern ranges.

Zygomorphic flowers

Alpine Sainfoin

Hedysarum hedysaroides
Legume family (Fabaceae)
JULY–AUG. 10–40CM

Note Flowers droop from elongated cluster up to 10cm long.
Description Leaves pinnate with 9–19 leaflets. Fruit capsule constricted between seeds, like a string of pearls.
Distribution Throughout the Alps, mainly on calcareous soils, to 2800m. Poor grassland, pastures. Scattered to rare.

Mountain Sainfoin

Onobrychis montana
Legume family (Fabaceae)
JUNE–AUG. 10–40CM

Note Calyx teeth 2–3 times longer than corolla tube.
Description Leaves pinnate with 7–15 narrowly-oval leaflets.
Distribution Mainly western Alps to 2000m, eastwards to Tyrol and Trentino. Meadows and pastures on dry, calcareous soils. Common.

Alpine Basil-thyme

Acinos alpinus
Labiate family (Lamiaceae)
JUNE–SEPT. 10–30CM

Note Whole plant smells of mint when crushed.
Description Flowers to 2.5cm long, hairy outside, with semicircular white markings inside. Leaves entire or with 1–3 teeth, with short hairs.
Distribution Calcareous regions to 2500m. Stony grassland, rocky debris. Scattered.

Zygomorphic flowers

Alpine Woundwort

Stachys alpina
Labiate family (Lamiaceae)
JULY–SEPT. 40–100CM

Note Flowers brown-red, hairy.
Description Lacking basal leaves. Stem slightly sticky. Stem leaves short-stalked and hairy.
Distribution Nutrient-rich calcareous soils between 600 and 2000m. Tall-herb communities, including those on soils enriched by livestock dung, open woodland. Scattered.

Alpine Betony

Stachys pradica
Labiate family (Lamiaceae)
JULY–SEPT. 10–30CM

Note Flowers in tightly packed flower-head.
Description Leaf rosette and 2–3 pairs of softly hairy stem leaves with crenate margins.
Distribution Nutrient-poor, dry soils, 1200–2500m. Mountain meadows, dwarf-shrub heaths. Rare.

Wall Germander

Teucrium chamaedrys
Labiate family (Lamiaceae)
JULY–AUG. 15–30CM

Note Flower lacks upper lip.
Description Flowers mostly grow to one side. Leaves with deep, blunt teeth. Leaves and stems softly hairy.
Distribution Nutrient-rich calcareous, warm soils between 600 and 1700m. Dry woods, stony grassland. Scattered.

Zygomorphic flowers

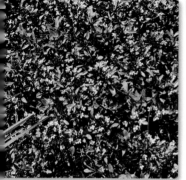

Hairy Thyme

Thymus praecox ssp. *polytrichus*
Labiate family (Lamiaceae)
JUNE–SEPT. 3–15CM

Note Plant has aromatic smell.
Description Mat-forming. Stems rectangular in cross-section, and hairy along two sides. Upper leaves larger than lower. Used as basis for cough medicines. Very variable species, with several similar species.
Distribution Dry soils to 3000m, especially common on calcareous substrates.

Kerner's Lousewort

Pedicularis kerneri
Figwort family (Scrophulariaceae)
JUNE–AUG. 5–15CM ▼ (☠)

Note Hairy, bending stems.
Description Upper lip with long beak, lower lip glabrous. Like all louseworts, semi-parasitic on other plants.
Distribution Central Alps, east to Hohe Tauern. Acid rocky debris and poor grassland between 1800 and 3000m.

Truncate Lousewort

Pedicularis recutita
Figwort family (Scrophulariaceae)
JUNE–AUG. 20–50CM ▼ (☠)

Note Flowerhead looks like it has been cut off at the top.
Description Flowers brownish-red, in cylindrical racemes. Upper lip lacking beak. Leaves pinnately-cut.
Distribution Damp, nutrient-rich and acid soils, 1200–2500m. Tall-herb communities, around springs and streams. Rare.

Zygomorphic flowers

Pink Lousewort

Pedicularis rosea
Figwort family (Scrophulariaceae)
JUNE–AUG. 5–15CM ▼

Note Upper stem has white-woolly hairs.
Description Flowers pink, upper lip without beak. Leaves pinnately-lobed, with leaflets widely spaced.
Distribution South-western Alps and northern and southern calcareous Alps to 2900m. Rock clefts and scree. Very rare.

Beaked Lousewort

Pedicularis rostratocapitata
Figwort family (Scrophulariaceae)
JUNE–AUG. 5–15CM ▼

Note Tight, rounded flowerheads.
Description Upper lip with beak. Lower lip with ciliated margins. Stem leaves almost opposite, glabrous. Leaves pinnately-lobed, leaflets pinnate.
Distribution Eastern Alps, mainly on calcareous soils, 1200–2500m. Stony soils, patchy grassland. Scattered.

Flesh-pink Lousewort

Pedicularis rostratospicata
Figwort family (Scrophulariaceae)
JULY–AUG. 20–40CM ▼

Note Inflorescence elongated spike-like.
Description Upper lip with beak. Lower lip glabrous. Stem leaves alternate. Leaves pinnately-lobed, leaflets cut.
Distribution Between 2000 and 2700m. Poor grassland, stony pastures, rocky debris. Scattered to rare.

Zygomorphic flowers

Whorled Lousewort

Pedicularis verticillata
Figwort family (Scrophulariaceae)
JUNE–AUG. 5–20CM ▼ (☠)

Note The only lousewort with whorls of 3–4 stem leaves.
Description Flowers in compact raceme. Lower lip of flower divided into three. Leaves pinnately-lobed. Basal leaves in rosette.
Distribution To 2800m on poor, mainly calcareous soils. Pastures and grassland. Scattered.

Heath Spotted-orchid

Dactylorhiza maculata
Orchid family (Orchidaceae)
JUNE–JULY 20–60CM ▼

Note Flowers pale pink with darker stripes and spots.
Description Inflorescence spike cylindrical. Flower lip three-lobed, with pointed middle lobe. Stem thin, solid or somewhat hollow. Leaves spotted.
Distribution Unfertilized periodically wet meadows and pastures to 2000m. Scattered. Similar species.

Broad-leaved Marsh-orchid

Dactylorhiza majalis
Orchid family (Orchidaceae)
MAY–JULY 20–60CM ▼

Note Leaves mostly with large, irregular brownish-red spots on upperside.
Description Inflorescence spike cylindrical. Flowers purple-red, with darker markings. Stem thick and hollow. Leaves 3–6 times as long as wide.
Distribution Wet soils, usually calcareous, to 2300m. Fens, wet meadows and pastures. Scattered.

Zygomorphic flowers

Early Marsh-orchid

Dactylorhiza incarnata ssp. *cruenta*
Orchid family (Orchidaceae)
JUNE–JULY 15–35CM ▼

Note Leaves with markings above and below.
Description Flowers pink (may be blood-red, especially in ssp. coccinea). NB complex species, with a number of subspecies. Outer tepals turning upwards. Flower lip with central tip. Leaves ovate and pointed.
Distribution Calcareous damp meadows and fens to 2500m. Rare.

Dark-red Helleborine

Epipactis atrorubens
Orchid family (Orchidaceae)
JUNE–JULY 20–50CM ▼

Note Flowers vanilla-scented.
Description Flowers red-brown with yellow centre (anther cap). Front of lip with two cusps. Leaves elongate-ovate, almost in two rows.
Distribution Mainly northern and southern ranges. Calcareous, dry soils to 2000m. Open woodland, Mountain Pine communities. Rare.

Fragrant Orchid

Gymnadenia conopsea
Orchid family (Orchidaceae)
MAY–AUG. 10–60CM ▼

Note Flower spike very long; flowers unmarked.
Description Flowers with long spurs, pale to dark pink (sometimes white), vanilla-scented. Lip lobes more or less equal in length. Leaves linear and broad, to 11mm wide.
Distribution Mostly calcareous meadows, to 2800m. Scattered to common.

Zygomorphic flowers

Short-spurred Fragrant Orchid

Gymnadenia odoratissima
Orchid family (Orchidaceae)
JUNE–AUG. 15–50CM ▼

Note Leaves narrow and grass-like.
Description Flowers short-spurred, pale to dark pink (sometimes white), vanilla-scented. Central lobe of lip long. Leaves to 7mm wide.
Distribution Only on calcareous, usually damp soils, to 2700m. Scattered in central Alps, otherwise very rare.

Black Vanilla Orchid

Gymnadenia rhellicani
Orchid family (Orchidaceae)
JUNE–AUG. 10–30CM ▼

Note Flowers dark red (look almost black from a distance).
Description Flowerheads dense, almost globular to pyramidal. Lip curved upwards. Leaves narrow, almost grass-like. Plant vanilla-scented.
Distribution To 2800m. Poor, mainly calcareous soils. Stony mountain meadows, grassland. Rare.

Red Vanilla Orchid

Gymnadenia rubra
Orchid family (Orchidaceae)
MAY–JULY 10–30CM ▼

Note Flowers uniformly bright red.
Description Flowerheads dense, oval to globular. Lip curved upwards. Leaves narrow, almost grass-like.
Distribution To 2600m. Poor, calcareous soils. Stony mountain meadows, grassland. Rare.

Zygomorphic flowers

Early-purple Orchid

Orchis mascula
Orchid family (Orchidaceae)
MAY–JULY 20–50CM ▼

Note Flowers in long, cylindrical somewhat loose spike.
Description Flowers with long, upward-pointing spur, red-violet (sometimes white), with some dark spots.
Distribution Nutrient-poor, dry soils to 2600m. Meadows, dry grassland, scrub. Scattered.

Burnt Orchid

Neotinea ustulata
Orchid family (Orchidaceae)
JUNE–JULY 10–30CM ▼

Note Flowerhead dark brownish-red or blackish towards the top.
Description Central lobe of lip divided, pale with red spots.
Distribution Nutrient-poor, usually dry, calcareous soils, to 2000m. Sunny meadows, dry grassland.

Round-headed Orchid

Traunsteinera globosa
Orchid family (Orchidaceae)
JUNE–JULY 20–50CM ▼

Note Slim-looking plant.
Description Flowerhead dense, rounded to oval. Lips with dark spots. Some of the petals with long, spoon-shaped tips. Leaves narrow and erect, blue-green.
Distribution Nutrient-poor, somewhat damp calcareous soils, to 2800m. Scattered.

Four petals

Alpine Poppy

Papaver alpinum ssp. *sendtneri*
Poppy family (Papaveraceae)
JULY–AUG. 5–15CM ▼

Note Stem with adpressed hairs.
Description Flowers to 5cm across (white in this subspecies). Basal leaves 1–2 pinnate, with pointed tips. Plant has white milky sap.
Distribution Northern Alps to 2600m, central Switzerland and eastwards. Calcareous rocks and scree. Related forms in west and southern Alps.

Alpine Rock-cress

Arabis alpina
Crucifer family (Brassicaceae)
MAY–SEPT. 10–40CM

Note All leaves roughly-hairy.
Description Grows in loose cushions. Flowers, to 1.5cm diameter, in umbel-like racemes. Leaves with blunt teeth, upper leaves encircling stem.
Distribution Stony, calcareous soils between 1200 and 3200m. Common in calcareous Alps, otherwise rare.

Dwarf Alpine Rock-cress

Arabis pumila
Crucifer family (Brassicaceae)
JUNE–AUG. 5–20CM

Note Flowerheads seem large compared with leaf rosettes.
Description Basal leaves roughly-hairy, edged with bifid hairs. Stem leaves stalkless, rounded at base. Pods flattened and winged.
Distribution Calcareous regions of Alps to 3000m. Rock clefts and scree. Scattered.

Four petals

Shiny Rock-cress

Arabis soyeri
Crucifer family (Brassicaceae)
JULY–AUG. 10–25CM

Note Leaves obviously shiny.
Description Lower leaves entire or with few teeth, almost glabrous, with at most a few hairs. Pods upright.
Distribution Wet, calcareous soils to 2800m. Spring-fed flushes and streamsides, wet rocks. Scattered.

Alpine Braya

Braya alpina
Crucifer family (Brassicaceae)
JUNE–AUG. 5–15CM ▼

Note Leaves long and narrow.
Description Flowers white to pink, turning violet when older. Stem hairy, often reddish. Fruit pod to 12mm long.
Distribution Only in central eastern Alps, to 3000m on dry, basic soils. Stony, patchy grassland, moraines. Rare.

Alpine Bitter-cress

Cardamine bellidifolia ssp. *alpina*
Crucifer family (Brassicaceae)
JULY–AUG. 3–12CM

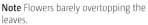

Note Flowers barely overtopping the leaves.
Description Leaves all undivided and mostly entire, or weakly-toothed.
Distribution Mainly central Alps on base-poor, damp soils. Snow-patches, spring-fed flushes, meadows. Rare on calcareous soils.

Four petals

Large Bitter-cress

Cardamine amara
Crucifer family (Brassicaceae)
MAY–JULY 10–50CM

Note The only bitter-cress with violet stamens.
Description Flowers, to 1.2cm diameter, in umbel-like raceme. Leaves pinnate, with 4–10 pairs of leaflets.
Distribution Throughout Alps to 2500m, on damp soils. Spring-fed flushes, streams and ditches. Scattered.

Mignonette-leaved Bitter-cress

Cardamine resedifolia
Crucifer family (Brassicaceae)
MAY–AUG. 3–15CM

Note Basal leaves form rosette. Lowest leaves entire, upper leaves 3-lobed.
Description Stem leaves mostly 5-lobed, with a pair of auricles encircling stem.
Distribution Stony, base-free soils between 1200 and 3200m. Scattered, rare in northern Alps.

Carinthian Whitlowgrass

Draba siliquosa
Crucifer family (Brassicaceae)
JUNE–AUG. 3–12CM ▼

Note Small rosette leaves are lanceolate and hairy.
Description Flowers small, 4–6mm diameter. Stems leafy, upper stems without hairs. Fruit glabrous.
Distribution To 2400m. Stony, basic, exposed soils. Rocks, crevices, scree. Scattered.

Four petals

Woolly Whitlowgrass

Draba tomentosa
Crucifer family (Brassicaceae)
JUNE–AUG. 3–8CM ▼

Note Leaves, stems and fruits all woolly-haired.
Description Flowers 7–10mm diameter. Basal leaves ovate, grey-green, in dense rosette.
Distribution Dry, calcareous soils to 3400m. Rock crevices, rubble, scree, often in wind-exposed sites. Rare.

Rock Candytuft

Iberis saxatilis
Crucifer family (Brassicaceae)
APRIL–JUNE 5–15CM

Note Outer petals much longer than inner petals.
Description Umbel-like flowerhead. Grows notably taller by fruiting stage. Leaves linear and sharply-pointed.
Distribution Only in south-western Alps, to 2000m. Stony grassland on dry, calcareous rocky soils. Rare.

Rock Kernera

Kernera saxatilis
Crucifer family (Brassicaceae)
JUNE–AUG. 10–40CM

Note Tall, slender, loose inflorescence arising from a small basal rosette.
Description Basal leaves and lower stem hairy. Fruits almost globular, 2–3mm broad, glabrous.
Distribution To 2700m on calcareous soils. Common in calcareous Alps, otherwise rare.

Four petals

Chamois Cress

Hornungia alpina
Crucifer family (Brassicaceae)
MAY–AUG. 3–15CM

Note Rosette leaves 5–11-divided, fleshy.
Description Dense umbel-like flower-heads, lengthening in fruit. Fruits 4–5mm, elongate-oval, pointed and flattened, glabrous.
Distribution Mainly on loose calcareous rocks, to 3200m. Scattered.

Alpine Penny-cress

Thlaspi caerulescens
Crucifer family (Brassicaceae)
APR–JUNE 15–40CM

Note Stem leaves at base clasping stem, with auricles.
Description Flowerheads rather tight, with violet anthers. Leaves blue-green, glabrous.
Distribution In the Alps to 2000m, mainly in the western Alps. Calcareous soils in meadows and pastures. Scattered to rare.

Alpine Bastard-toadflax

Thesium alpinum
Sandalwood family (Santalaceae)
JUNE–JULY 10–25CM

Note Flowers mostly 4-partite, look green from outside.
Description Flowers droop to one side above a whorl of bracts. Fruiting stems become upright. Leaves long and narrow.
Distribution Nutrient-rich calcareous soils to 3000m. Open woods, scrub, dry grassland. Scattered.

Four petals

Alpine Daphne

Daphne alpina
Daphne family (Thymelaeaceae)
MAY–JUNE 10–50CM ▼ ☠

Note Small shrub, smelling of vanilla.
Description Flowers short-stalked, with pointed petals, hairy on outside. Leaves in bushy bunches at ends of stems, grey-green above. Bark hairy and wrinkled. Berries red, poisonous!
Distribution Warm, dry, stony soils on limestone, to 2000m. Rare.

Alpine Bedstraw

Galium anisophyllon
Bedstraw family (Rubiaceae)
JULY–SEPT. 5–20CM

Note Umbel-like flowerheads, longer than whorled leaves.
Description Leaves in whorls of (7–)8 on rectangular stem. Non-spreading.
Distribution Nutrient-poor, dry, calcareous soils between 1200 and 2500m. Meadows and pastures, rocky debris. Scattered.

Swiss Bedstraw

Galium megalospermum
Bedstraw family (Rubiaceae)
JULY–SEPT. 3–10CM

Note Freely spreading by runners close to the ground.
Description Flowerheads, about same length as whorled leaves. Flowers yellowish-white. Leaves in whorls of (6–)7 on rectangular stem.
Distribution Stony calcareous soils, 1800–2600m. Rocky debris. Scattered.

Five petals

Aconite-leaved Buttercup

Ranunculus aconitifolius
Buttercup family (Ranunculaceae)
MAY–JULY 20–70CM ▼ ☠

Note The palmately-lobed leaves resemble those of Monkshood.
Description Flowers stalked. Stem hairy below the flowers. Leaves 3–5-palmate, with free central lobe.
Distribution On damp, nutrient-rich soils between 600 and 2500m. Banks, meadows and pastures. Common.

Large White Buttercup

Ranunculus platanifolius
Buttercup family (Ranunculaceae)
JUNE–JULY 40–120CM ▼ ☠

Note Central lobe not free to the base. Similar to previous species, but larger.
Description Flowers long-stalked. Stem below flowers glabrous. Leaves 5–7-partite.
Distribution Damp, nutrient-rich soils in woods and tall-herb communities, between 1200 and 2000m. Scattered.

Glacier Buttercup (Glacier Crowfoot)

Ranunculus glacialis
Buttercup family (Ranunculaceae)
JULY–AUG. 5–20CM ▼ ☠

Note Flowers long-lasting. Petals often turn pink.
Description Flowers to 3cm diameter. Sepals reddish-brown and hairy. Leaves fleshy, basal leaves deeply-lobed.
Distribution One of Europe's highest altitude flowers, to over 4200m.

Five petals

Alpine Buttercup

Ranunculus alpestris
Buttercup family (Ranunculaceae)
JUNE–AUG. 5–10CM ▼ ☠

Note Basal leaves richly shiny.
Description Flowers solitary or in pairs at end of stem. Basal leaves rounded, 3–5-lobed.
Distribution Mainly calcareous soils of outer ranges, between 1700 and 2900m. Damp, stony soils, snow-patches, open grassland. Scattered.

Parnassus-leaved Buttercup

Ranunculus parnassifolius
Buttercup family (Ranunculaceae)
JUNE–AUG. 5–20CM ▼ ☠

Note Leaves are rounded-heart-shaped.
Description Flowers to 2.5cm diameter, becoming reddish. Upper stem with felty hairs. Leaves blue-green.
Distribution Damp, calcareous soils between 1800 and 3000m. Fine rocky debris, scree. Rare.

Kuepfer's Buttercup

Ranunculus kuepferi
Buttercup family (Ranunculaceae)
JUNE–AUG. 5–20CM ▼ ☠

Note Leaves narrowly-lanceolate, almost grass-like.
Description Flowers usually solitary, to 3cm across. Leaves parallel-veined, entire and glabrous.
Distribution Mainly central and southern Alps to 3000m. Damp, acid soils; pastures and grassland. Scattered; rare in the north.

Five petals

Fringed Sandwort

Arenaria ciliata
Pink family (Caryophyllaceae)
JULY–AUG. 3–10CM

Note Flowers 3–5 on the stem.
Description Grows in loose mats. Flowers to 1cm diameter. Leaves pointed, 3–4 times as long as wide, with hairy edges.
Distribution To over 3000m on damp, stony, calcareous sites. Grassland and rocky debris. Common.

Alpine Mouse-ear

Cerastium alpinum
Pink family (Caryophyllaceae)
JULY–AUG. 5–15CM

Note Plant covered with soft hairs.
Description Flowers 1–3 on the stem, short-stalked, petals notched, calyx with glandular hairs. Leaves about 1cm long, grey-green.
Distribution Mainly in western Alps on acid soils. Stony grassland, rocky debris and scree. Scattered.

Field Mouse-ear

Cerastium arvense
Pink family (Caryophyllaceae)
MAY–AUG. 5–15CM

Note Flowers rather large – to 22mm across.
Description Petals noticeably deeply-notched. Leaves very narrow, pointed, green.
Distribution Throughout Alps to 2800m. Meadows and pastures, rocky slopes, walls. Common.

Five petals

Starwort Mouse-ear

Cerastium cerastioides
Pink family (Caryophyllaceae)
JULY–AUG. 5–15CM

Note Leaves glabrous.
Description Creeping growth, with upright flowering stems. Leaves fleshy, to 1cm long.
Distribution To 2800m on damp, nutrient-rich sites with long snow-cover. Snow-patches, spring-fed flushes. Common.

Broad-leaved Mouse-ear

Cerastium latifolium
Pink family (Caryophyllaceae)
JULY–AUG. 5–10CM

Note Leaves pointed, with short hairs, broadest in the middle.
Description Flowers 1–3 on the stem, long-stalked. Petals only slightly notched. Leaves ovate-lanceolate.
Distribution Between 1700 and 3400m on damp calcareous sites with little fine soil. Scattered.

Glacier Mouse-ear

Cerastium uniflorum
Pink family (Caryophyllaceae)
JULY–AUG. 3–15CM

Note Flowers 3–5 on the stem – almost umbel-like.
Description Forms dense mats. Leaves about 1.5cm long, blunt, upper leaves broadest.
Distribution Between 2000 and 3400m on acid soils. Rocky grassland, scree and gravel. Common in central Alps, otherwise rare.

Five petals

Alpine Gypsophila

Gypsophila repens
Pink family (Caryophyllaceae)
MAY–AUG. 7–25CM

Note Petals notched.
Description Produces many non-flowering shoots. Flowers in loose panicle, white to pink. Leaves linear-lanceolate, grey-green, glabrous.
Distribution Throughout Alps on limestone to 3000m. Stony soils, debris, scree and rock. Common.

Spring Sandwort

Minuartia verna
Pink family (Caryophyllaceae)
MAY–AUG. 5–15CM ▼

Note Leaves are needle-like.
Description Forms loose mats. Flowers to 1cm diameter, petals rounded, sepals 3-nerved.
Distribution Throughout Alps on damp calcareous soils to 3200m. Stony grassland, rocky debris, gravel. Common.

Creeping (Ciliate) Sandwort

Moehringia ciliata
Pink family (Caryophyllaceae)
JUNE–AUG. 3–20CM

Note Needle-shaped leaves fringed with ciliate hairs.
Description Creeping growth. Flowers 1–3 on the stem. Petals narrow. Leaves pointed and fleshy.
Distribution Throughout Alps, mainly on limestone, to over 3000m. Damp, rocky soils. Further similar species. Common.

Five petals

Alpine Pearlwort

Sagina saginoides
Pink family (Caryophyllaceae)
JUNE–AUG. 2–8CM

Note Petals are shorter than the sepals.
Description Flowers small, to 8mm diameter. Leaves narrow and pointed, to 20mm long.
Distribution Throughout Alps, on damp, acid soils to 3200m. Snow-patches, spring-fed flushes, poor grassland. Scattered to common.

Small Campion

Silene pusilla
Pink family (Caryophyllaceae)
JULY–SEPT. 5–20CM

Note Petal tips often have four small teeth.
Description Flowers long-stalked. Stems branched, sticky towards the top. Leaves 1–2mm broad, to 3cm long.
Distribution Limestone regions between 800 and 2400m. Damp or wet soils, rocky sites, scree. Scattered.

Rock Campion

Silene rupestris
Pink family (Caryophyllaceae)
JUNE–AUG. 10–25CM ▼

Note Leaves blue-green, glabrous.
Description Loose branching flowering stems with few flowers. Petals notched. Leaves to 2cm long.
Distribution Central and southern Alps on acid soils to 2600m. Rocky slopes, stony grassland, rock crevices. Scattered.

Five petals

Bladder Campion

Silene vulgaris
Pink family (Caryophyllaceae)
JUNE–SEPT. 20–50CM

Note Calyx inflated, bladder-like.
Description Flowers to 22mm diameter. Petals deeply-notched, white, rarely pink. Leaves glabrous, or sometimes softly hairy.
Distribution From lowland to 2800m. Dry, poor soils. Meadows, scree. Common.

Two-flowered Sandwort

Arenaria biflora
Pink family (Caryophyllaceae)
JULY–AUG. 1–25CM

Note Stems 1–2-flowered.
Description Plant spreading. Flowers with 3 styles, as in all sandworts. Leaves rounded, hairy only on stalks.
Distribution Mainly on central ranges of the Alps, between 1700 and 3100m. Acid soils with long snow-cover. Common.

Nottingham Catchfly

Silene nutans
Pink family (Caryophyllaceae)
JUNE–AUG. 20–60CM

Note Upper stem sticky.
Description Flowers droop to one side. Petals often rolled upwards, deeply cleft.
Distribution To 2400m. Dry, nutrient-poor soils. Open woods, dry meadows. Scattered to common.

Five petals

Alpine Knotgrass

Polygonum alpinum
Dock family (Polygonaceae)
JUNE–JULY 30–60CM

Note Flowers in a loose, almost pyramidal panicle.
Description Flowers creamy yellow to pink. Leaves lanceolate, to 15cm long, paler beneath.
Distribution Mainly western and southern Alps, to 2200m. Nutrient-rich base-poor soils. Meadows, scrub. Scattered to rare.

Alpine Bistort

Persicaria vivipara
Dock family (Polygonaceae)
JUNE–AUG. 5–25CM

Note Viviparous reproduction via bulbils that develop in the lower part of the inflorescence.
Description Flowers pale pink or white. Leaves to 8cm long, narrowly-lanceolate, grey-green below, with rolled edges.
Distribution Nutrient-poor soils to 3000m. Meadows, pasture, fens. Common.

Mountain Bearberry

Arctostaphylos alpinus
Heath family (Ericaceae)
MAY–JUNE 5–20CM

Note Deciduous dwarf shrub. Leaves have hairy margins.
Description Flowers bell-shaped, in groups of 2–5, pale green to pink. Leaves thin and finely-toothed. Berries black when ripe.
Distribution On humus-rich soils between 1800 and 2600m. Moorland, open coniferous woods, Mountain Pine scrub. Scattered.

Five petals

Bearberry

Arctostaphylos uva-ursi
Heath family (Ericaceae)
MAY–JULY 5–15CM ▼ (☠)

Note Evergreen dwarf-shrub. Forms cushions and has leathery leaves.
Description Flowers bell-shaped, drooping in groups of 3–10, white to pink. Leaves flat, dark green and entire. Berries red when ripe. Ancient medicinal plant.
Distribution 600–2700m. Dry, nutrient-poor soils. Moorland, pastures. Common.

One-flowered Wintergreen

Moneses uniflora
Heath family (Ericaceae)
JUNE–AUG. 5–15CM ▼

Note Flower and stalk resemble a tiny umbrella.
Description Flowers solitary, to 2cm across. Leaves rounded and toothed in a rosette. Fruiting capsule upright.
Distribution Nutrient-poor, humus- and moss-rich acid soils. Coniferous woodland, between 1200 and 2000m. Scattered.

Serrated Wintergreen

Orthilia secunda
Heath family (Ericaceae)
JUNE–JULY 10–25CM ▼

Note Bell-shaped flowers droop to one side.
Description Flowers bell-shaped to globose, greenish-white. Leaves on lowest third of stem weakly-toothed.
Distribution Humus- and moss-rich woodland and dwarf-shrub heath between 600 and 2300m. Rare.

Five petals

Round-leaved Wintergreen

Pyrola rotundiflora
Heath family (Ericaceae)
JUNE–SEPT. 10–30CM ▼

Note Unlike previous species, the flowers do not droop to one side.
Description Flowers 8–30, nodding, open bell-shaped. Stem green. Leaves round, 2–3cm long, often heart-shaped at base.
Distribution Acid, humus-rich woods between 600 and 2400m. Scattered to common.

Bog Bilberry

Vaccinium uliginosum ssp. *pubescens*
Heath family (Ericaceae)
JUNE–JULY 5–20CM

Note This subspecies has a prostrate growth habit.
Description Flowers white to pink, mostly solitary. Leaves paler beneath. Fruit dark blue when ripe.
Distribution Between 1500 and 2700m. Acid, humus-rich soils. Rocky sites, dwarf-shrub heath. Scattered.

Cowberry

Vaccinium vitis-idaea
Heath family (Ericaceae)
JUNE–JULY 5–25CM

Note Leaves have rolled edges and dark glands on underside.
Description Flowers white or pink. Leaves evergreen, leathery, shiny – used as medicinal tea. Berries red and edible.
Distribution Acid soils in woods and moorland, to 2700m.

Five petals

Sweetflower Rock-jasmine

Androsace chamaejasme
Primrose family (Primulaceae)
JUNE–JULY 3–12CM ▼

Note Long stem covered with villous hairs.
Description Loose basal leaf rosettes. Petals rounded. Leaves to 1.5cm long, with hairy margins.
Distribution To 3000m, mainly in northern Alps. Stony, calcareous sites. Grassland and rocky debris. Scattered.

Milkwhite Rock-jasmine

Androsace lactea
Primrose family (Primulaceae)
JUNE–JULY 3–20CM ▼

Note Rosette leaves very narrow – at most 2mm broad and 2cm long.
Description Petals emarginate (shallow notch at tip). Stem glabrous, leaves only hairy at tip.
Distribution Northern Alps. Stony, calcareous sites to 2300m. Grassland and rocks. Rare.

Hairy Rock-jasmine

Androsace villosa
Primrose family (Primulaceae)
JUNE–JULY 2–5CM ▼

Note Stem and undersides and margins of leaves with dense covering of long hairs.
Description Flowers to 1cm diameter. Petals rounded. Leaves in almost globular rosettes, glabrous above.
Distribution Calcareous stony soils in southern Alps between 1200 and 3000m. Rare.

Five petals

Dolomite Rock-jasmine

Androsace hausmannii
Primrose family (Primulaceae)
JULY–AUG. 1–4CM ▼

Note Often retains remains of previous year's flowers.
Description Flowers solitary on stems to 12mm long. Petals slightly notched. Leaves roughly-hairy, in loose rosettes.
Distribution Eastern Alps, only on limestone, to 3000m. Rock crevices and scree. Grassland. Rare.

Blunt-leaved Rock-jasmine

Androsace obtusifolia
Primrose family (Primulaceae)
JULY–AUG. 5–12CM ▼

Note Often retains remains of previous year's flowers.
Description Petals rounded. Stem often reddish, with very short hairs. Leaves with marginal ciliate hairs.
Distribution Poor, humus-rich soils to 3000m in central Alpine ranges. Rock crevices, grassland, pastures. Scattered.

Swiss Rock-jasmine

Androsace helvetica
Primrose family (Primulaceae)
MAY–JULY 1–3CM ▼

Note Flowers grow flat against the thick, firm cushions of leaf rosettes.
Description Petals rounded. Flower stalks shorter than the leaves. Leaves grey-green.
Distribution Mainly in northern Alps. Stony, calcareous sites between 1500 and 3600m. Scattered to rare.

Five petals

White Stonecrop

Sedum album
Stonecrop family (Crassulaceae)
JUNE–SEPT. 5–25CM

Note Long-stalked, umbel-like inflorescence. Long, leafy stem often reddish.
Description Flowers white, petals with red central vein. Stem glabrous. Leaves cylindrical, fleshy and glabrous.
Distribution Thin soils, rock debris, rock crevices, scree, walls. Common.

Thick-leaved Stonecrop

Sedum dasyphyllum
Stonecrop family (Crassulaceae)
JUNE–AUG. 2–8CM

Note Flowers with 5 to 7 petals on the same plant.
Description Flowers white, often pink on outside; petals with red central vein. Leaves blue-green, often with red markings, fleshy, flattened above.
Distribution Acid, stony soils to 2500m. Rock crevices and gravel. Scattered to common.

Pyramidal Saxifrage

Saxifraga cotyledon
Saxifrage family (Saxifragaceae)
JUNE–AUG. 20–70CM ▼

Note Impressive panicles of up to 100 flowers.
Description Stem branching from base or below half way. Flowering plants with leaf rosettes to 15cm across.
Distribution South-western Alps on siliceous soils, east to Vorarlberg. Rock crevices between 900 and 2500m. Rare.

Five petals

Live-long Saxifrage

Saxifraga paniculata
Saxifrage family (Saxifragaceae)
MAY–AUG. 10–30CM ▼

Note Petals usually with many red spots on inside.
Description Plant spreading. Flowers in panicles, branches with 1–5 flowers. Basal leaves in rosettes. Leaves toothed, lime-encrusted around edges.
Distribution Dry, calcareous, stony soils, to 3400m. Scattered.

Host's Saxifrage

Saxifraga hostii
Saxifrage family (Saxifragaceae)
MAY–JULY 20–60CM ▼

Note Rosette leaves over 4cm long.
Description Stem branched in upper third, branches with 5–10 flowers. Rosette leaves linear, toothed, lime-encrusted around edges.
Distribution Two subspecies in southern calcareous Alps to 2500m. Rocks, stony slopes and rocky debris. Rare.

Encrusted Saxifrage

Saxifraga crustata
Saxifrage family (Saxifragaceae)
JUNE–AUG. 10–30CM ▼

Note Most leaves almost entirely encrusted with lime.
Description Stem branched in upper third, branches with 1–3 flowers. Rosette leaves mostly entire, to 4mm broad and over 2cm long.
Distribution Only in southern calcareous Alps to 2500m. Rocks, stony slopes and rock debris. Rare.

Five petals

Blue-grey Saxifrage

Saxifraga caesia
Saxifrage family (Saxifragaceae)
JULY–SEPT. 3–12CM ▼

Note Dense, firm bluish-green cushions are often encrusted with lime.
Description Inflorescence a raceme with 2–5 white flowers. Stem with few small, somewhat sticky leaves.
Distribution Calcareous stony soils to 3000m. Common.

Burser's Saxifrage

Saxifraga burseriana
Saxifrage family (Saxifragaceae)
APRIL–JULY 3–10CM ▼

Note Triangular, prickly rosette leaves resemble conifer needles.
Description Flowers to 15mm diameter, solitary. Stem and stem leaves sticky. Rosette leaves grey-green, to 12mm long.
Distribution Eastern Alps on calcareous stony soils. Rock crevices and scree. Rare.

Rough Saxifrage

Saxifraga aspera
Saxifrage family (Saxifragaceae)
JULY–AUG. 5–20CM ▼

Note White flowers, with yellow-violet spots inside.
Description Does not form basal rosettes. Non-flowering shoots with small rosettes in leaf axils. Leaves mostly with stiff ciliate hairs.
Distribution Acid, stony soils to 2500m. Rock crevices and scree. Scattered to rare.

Five petals

Cobweb Saxifrage

Saxifraga arachnoidea
Saxifrage family (Saxifragaceae)
JUNE–AUG. 10–30CM ▼

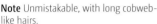

Note Unmistakable, with long cobweb-like hairs.
Description Flowers white to yellow-green, with yellow spots inside. Non rosette-forming.
Distribution Only in the southern Alps between Lake Garda and Lake Como. Sheltered rocky clefts to 1850m. Rare.

Scree Saxifrage

Saxifraga androsacea
Saxifrage family (Saxifragaceae)
MAY–AUG. 2–10CM ▼

Note Stem and undersides and margins of leaves covered in glandular hairs.
Description Petals pure white, 1–2 per stem. Leaves in rosette, stem leaves few or absent
Distribution Damp, mainly calcareous, stony soils to 3400m. Rocky ground, snow-patches. Common.

Round-leaved Saxifrage

Saxifraga rotundifolia
Saxifrage family (Saxifragaceae)
JUNE–SEPT. 10–60CM ▼

Note White flowers with red and yellow spots on upper side.
Description Inflorescence many-flowered. Basal leaves round–kidney-shaped, crenate–toothed, hairy.
Distribution Damp, nutrient-rich, calcareous soils to 2200m. Tall-herb communities, spring-fed flushes and streams, woods. Common.

Five petals

Starry Saxifrage

Saxifraga stellaris
Saxifrage family (Saxifragaceae)
JUNE–AUG. 5–20CM ▼

Note Petals narrow, each with two yellow spots.
Description Spreads by stolons. Stem leafless, rosette leaves fleshy, often coarsely-toothed towards the tip.
Distribution Throughout the Alps to over 3000m. Damp soils near springs and streams. Common.

Shield-leaved Saxifrage

Saxifraga cuneifolia
Saxifrage family (Saxifragaceae)
JUNE–JULY 10–25CM ▼

Note Leaves shield-shaped.
Description Spreads by stolons. Sepals reflexed. Lacks stem leaves. Rosette leaves broadly-oval, fleshy, weakly-toothed.
Distribution Mainly acid soils on damp, shady rocks in woodland and scrub. Scattered to rare.

Flat-leaved Saxifrage

Saxifraga muscoides
Saxifrage family (Saxifragaceae)
JUNE–AUG. 1–8CM ▼

Note Thick, mossy cushion formed from many tiny rosettes.
Description Flowers whitish to pale yellow, 1–2 on densely-hairy stems growing well clear above cushion. Plant smells resinous.
Distribution Acid, stony soils between 2000 and 4200m. Rock crevices and scree. Rare.

Five petals

Grass-of-Parnassus

Parnassia palustris
Saxifrage family (Saxifragaceae)
JULY–SEPT. 5–30CM ▼

Note Heart-shaped leaf on lower third of each stem.
Description Flowers solitary, 1–3.5cm diameter, yellow-green within. Basal leaves in rosette, entire, heart-shaped and stalked.
Distribution Throughout Alps, to over 2500m on wet, mainly calcareous soils. Common.

Clusius' Cinquefoil

Potentilla clusiana
Rose family (Rosaceae)
JUNE–AUG. 5–10CM ▼

Note Stem much longer than basal leaves.
Description Flowers to 2.5cm diameter. Stamens often reddish. Leaves palmate, with softly-hairy margins.
Distribution Only in eastern Alps, especially in northern and southern calcareous Alps. Common.

Shrubby White Cinquefoil

Potentilla caulescens
Rose family (Rosaceae)
JUNE–SEPT. 10–30CM

Note Undersides of leaves covered with long, silky hairs.
Description Flowers to 2.5cm diameter. Sepals visible between petals. Stamens form a kind of inner 'corolla'. Leaves palmate.
Distribution Mainly in rock crevices in calcareous regions, to 2000m. Common.

Five petals

Pyrenean Bastard-toadflax

Thesium pyrenaicum
Sandalwood family (Santalaceae)
JUNE–AUG. 10–50CM

Note Flowers mostly 5-partite and green on the outside (see p. 56, bottom).
Description Flowers in many-sided raceme. Leaves yellow-green, narrow, linear-lanceolate.
Distribution Acid, loamy soils to 2500m. Meadows, pastures, scrub. Scattered.

Alpine Wood Crane's-bill

Geranium rivulare
Crane's-bill family (Geraniaceae)
JUNE–AUG. 20–50CM ▼

Note White flowers with red veins.
Description Flowers 2–3 cm diameter. Stem angularly-branched, with 8–10 flowers. Leaves palmate, with cut lobes.
Distribution Mainly acid, humus-rich soils. Poor grassland, open woods, dwarf-shrub heath. Scattered.

Great Masterwort

Astrantia major
Umbellifer family (Apiaceae)
JUNE–AUG. 30–100CM ▼

Note The umbels, 2–4cm across are surrounded by a ring of white to pink bracts.
Description Bracts have crosswise veins. Lower leaves 5–7-palmately-lobed, the largest lobes not split to the base.
Distribution Nutrient-rich, usually calcareous soils to 2000m. Meadows and tall-herb communities. Common.

Five petals

Lesser Masterwort

Astrantia minor
Umbellifer family (Apiaceae)
JULY–AUG. 10–40CM ▼

Note Umbels pure white, 1.5cm diameter maximum.
Description Bracts lack crosswise veins. Lower leaves 5–7-lobed, the narrow lobes cut almost to the base.
Distribution South-western Alps, east to south Tyrol. Nutrient-poor soils to 2500m. Scattered.

Haller's Sermountain

Laserpitium halleri
Umbellifer family (Apiaceae)
JULY–AUG. 15–60CM ▼

Note Basal leaves to 50cm long and 4–5-pinnate.
Description Flowers in domed umbels. Bracts with 3 teeth at the tip and with ciliate hairs.
Distribution Central Alps to 2700m, east to Tyrol. Dry, nutrient-poor, acid soils. Meadows and Alpine grassland. Scattered.

Broad-leaved Sermountain

Laserpitium latifolium
Umbellifer family (Apiaceae)
JULY–AUG. 50–200CM ▼

Note Basal leaves to 1m long, 1–2-pinnate, hairy beneath.
Description Bracts and bracteoles glabrous. Leaves triangular in outline, lobes narrowly-lanceolate and coarsely-toothed.
Distribution To 2400m. On nutrient-poor soils subject to drying. Scattered.

Five petals

Narrow-leaved Sermountain

Laserpitium siler
Umbellifer family (Apiaceae)
JUNE–JULY 30–150CM

Note Basal leaves to 1m long, 3–4-pinnate.
Description Bracts and bracteoles glabrous. Leaves triangular in outline, lobes narrowly-lanceolate and entire.
Distribution Dry, calcareous, stony soils, between 600 and 2000m. Scattered.

Austrian Hogweed

Heracleum austriacum
Umbellifer family (Apiaceae)
JULY–SEPT. 10–60CM

Note Outer petals of outer flowers elongated.
Description Flowers white to pink. Stem smooth and thin (at base < 4mm across), upper stem softly-hairy. Leaves 1-pinnate, with 2–3 pairs of lobes.
Distribution Except for a locality in Switzerland, only in north-eastern Alps. Rare.

Hogweed

Heracleum sphondylium
Umbellifer family (Apiaceae)
JUNE–SEPT. 80–150CM

Note Outer flowers of umbel enlarged, with notched petals.
Description Flowers white, rarely pink. Stem thick, grooved and with stiff hairs. Leaves very large, 1-pinnate, with lobed, toothed segments.
Distribution Nutrient-rich soils, to 2400m. Meadows and tall-herb communities.

Five petals

Wild Angelica

Angelica sylvestris
Umbellifer family (Apiaceae)
JULY–SEPT. 50–200CM

Note Umbel consists of many globular smaller umbels.
Description Upper stem furrowed and hairy. Leaves triangular in outline, 2–3-pinnate.
Distribution Damp, nutrient-rich soils to 1800m. Woods and meadows. Common.

Masterwort

Peucedanum ostruthium
Umbellifer family (Apiaceae)
JUNE–AUG. 30–100CM

Note Leaves 1–2-ternate with oval, stalked, irregularly-toothed lobes.
Description Flowers white, rarely pink. Umbels domed, becoming funnel-shaped in fruit. Stem furrowed and hollow. Medicinal plant, also used for flavourings (e.g. spirits).
Distribution Nutrient-rich soils between 1400 and 2700m. Common.

Candy Carrot

Athamanta cretensis
Umbellifer family (Apiaceae)
MAY–AUG. 10–50CM

Note Petals hairy on the outside.
Description Umbels 6–12-rayed. Stem grey-green, hairy, leafless. Leaves 3-pinnate. Leaf tip to 1cm long and < 1mm wide. Fruit very hairy. Medicinal plant.
Distribution Stony, calcareous soils between 900 and 2500m. Rocky debris. Scattered.

Five petals

Caraway

Carum carvi
Umbellifer family (Apiaceae)
MAY–AUG. 20–60CM

Note Inflorescence much-branched, even from lower half of the stem.
Description Flowers white to pink. Plant glabrous. Leaves 2–3-pinnate, leaflets linear and deeply divided. Medicinal and culinary plant.
Distribution Meadows and pastures between 400 and 2200m. Common.

Spignel

Meum athamanticum
Umbellifer family (Apiaceae)
MAY–SEPT. 20–60CM

Note Whole plant strongly aromatic.
Description Flowers white, glabrous. Umbels 3–15-rayed. Stem grooved, almost leafless. Basal leaves 2–4-pinnate, soft, with finely-pointed tips. Medicinal and culinary plant, also used as flavouring (e.g. spirits).
Distribution Acid meadows and pastures to 2800m. Scattered.

Moon Carrot

Seseli libanotis
Umbellifer family (Apiaceae)
JULY–SEPT. 20–120CM

Note Stem almost glabrous, deeply-furrowed.
Description Umbels domed. Leaves grey-green, 2–3-pinnate. Lower leaflets often overlapping close to the leaf stalk. Fruits hairy and ribbed.
Distribution Mainly on calcareous grassland, between 400 and 2500m. Rare.

Five petals

Alpine Chervil

Chaerophyllum villarsii
Umbellifer family (Apiaceae)
MAY–SEPT. 30–120CM

Note Stems and undersides of leaves bristly-hairy.
Description Petals hairy. Leaves pinnate, segments often overlapping.
Distribution 1200–2400m, on periodically dry soils. Open woodland, meadows, tall-herb communities. Common.

Bogbean

Menyanthes trifoliata
Bogbean family (Menyanthaceae)
MAY–JUNE 15–30 cm ▼ (☠)

Note Petals with beard-like fringe on inner surface.
Description Flowers 5–20, in dense raceme, 2–3cm diameter. Leaves long-stalked and triangular, entire and somewhat fleshy.
Distribution Acid, nutrient-poor soils to 2300m. Mires and margins of lakes. Scattered.

Styrian Gentian

Gentiana frigida
Gentian family (Gentianaceae)
JUNE–SEPT. 5–15CM ▼

Note The almost white flowers make this species something of a speciality.
Description Flowers 1–3 on the stem, short-stalked. Petals with blue stripes on outside. Leaves linear-lanceolate.
Distribution Only in eastern central Alps to 2500m. Acid soils. Rare.

Five petals

Common Vincetoxicum

Vincetoxicum hirundinarea
Milkweed family (Asclepiadaceae)
JUNE–AUG. 30–100CM ▼ ☠

Note Yellowish-white flowers in upper axillary inflorescence.
Description Corolla funnel-shaped. Leaves opposite, entire, heart-shaped and pointed.
Distribution Stony, usually calcareous soils between 600 and 2500m. Open woodland and scree. Scattered.

White Mullein

Verbascum lychnitis
Figwort family (Scrophulariaceae)
JUNE–OCT. 50–130CM

Note Flowers grouped in tall, branching, mealy inflorescence.
Description Flowers white or yellow. Stamens with villous hairs. Stem angular. Underside of leaves grey-green and hairy.
Distribution Calcareous stony soils to 2200m. Scree, roadsides and woodland edges. Common.

Twinflower

Linnaea borealis
Honeysuckle family (Caprifoliaceae)
JULY–AUG. 5–15CM ▼

Note Twin 5-lobed flowers at the top of stem.
Description Creeping growth, to over 1m. Flower pink inside, vanilla-scented. Leaves opposite, leathery, weakly-toothed.
Distribution Moss-rich, acid soils in shady coniferous forest, especially in central Alps. Rare.

Five petals

Rock Valerian

Valeriana saxatilis
Valerian family (Valerianaceae)
JUNE–AUG. 5–30CM

Note Only one pair of leaves per stem.
Description Flowers clustered. Stem leaves opposite, broadly-lanceolate and entire, with 3 parallel main veins.
Distribution Calcareous stony soils to 2500m. Scree, poor grassland and gravel. Rare.

Three-leaved Valerian

Valeriana tripteris
Valerian family (Valerianaceae)
APRIL–JULY 10–50CM

Note Stems with 2–5 pairs of leaves with three unequal lobes.
Description Flowers pale pink, in umbel-like clusters. In fruit, the calyx forms a hairy tuft.
Distribution Moderately damp, usually calcareous, stony soils to 2500m. Rock crevices and rocky debris.

Scree Valerian

Valeriana saliunca
Valerian family (Valerianaceae)
JULY–AUG. 3–10CM ▼

Note Pale pink flowers in tight flowerheads.
Description Flowers closely surrounded by reddish bracts. Flowering stems with a single pair of leaves, glabrous.
Distribution Western Alps, east to northern and southern Tyrol, between 1800 and 2700m. Rock crevices and poor grassland. Rare.

Many petals

Tyrol Anemone

Anemone baldensis
Buttercup family (Ranunculaceae)
JUNE–AUG. 5–15CM ▼ ☠

Note Petals 6–8, hairy on outside.
Description Flowers solitary, 2.5–4cm diameter. Basal leaves stalked and biternate. Fruit hairy.
Distribution Calcareous soils to 3100m. Scree and stony grassland. Scattered in southern Alps, rare in northern Alps.

Narcissus-flowered Anemone

Anemone narcissiflora
Buttercup family (Ranunculaceae)
MAY–JULY 20–50CM ▼ ☠

Note Flowers 3–8 in terminal umbel.
Description Flowers 2–3cm diameter. Basal leaves stalked, 3–5-palmately-lobed. Fruit glabrous.
Distribution Mainly in northern and southern mountain ranges, between 800 and 2500m. Calciphile. Scattered.

Coriander-leaved Callianthemum

Callianthemum coriandrifolium
Buttercup family (Ranunculaceae)
MAY–JULY 10–30CM ▼ ☠

Note Leaves resemble those of Coriander.
Description Flowers to 3cm diameter; petals 6–12. Leaves glabrous, blue-green, 1–2-pinnate.
Distribution Central and southern Alps between 1600 and 2800m. Mountain meadows and pastures. Rare.

Many petals

Christmas Rose

Helleborus niger
Buttercup family (Ranunculaceae)
DEC.–MAY 10–25CM ▼ ☠

Note One of few species that flower in winter. Also known as Black Hellebore.
Description Flowers solitary, 6–10cm diameter, white to pink. Basal leaves leathery, with 8–9 toothed segments.
Distribution North-eastern and southern calcareous Alps between 500 and 1800m. Open woodland. Often cultivated.

Alpine Pasqueflower

Pulsatilla alpina ssp. *alpina*
Buttercup family (Ranunculaceae)
APRIL–JUNE 15–55CM ▼ ☠

Note Flowers nodding at first, becoming upright in full bloom.
Description Flowers 4–6cm diameter, somewhat bluish on outside. Leaves whorled on stem.
Distribution Northern and southern Alpine ranges, between 1500 and 2600m. Calcareous stony soils. Meadows and pastures. Scattered.

Spring Pasqueflower

Pulsatilla vernalis
Buttercup family (Ranunculaceae)
APRIL–JUNE 5–20CM ▼ ☠

Note Outside of flower and bracts with golden hairs.
Description Flowers 2–4cm diameter, pink on outside with bluish or violet tint. Lacking stem leaves.
Distribution Acid soils from 1400 to over 3000m. Open coniferous woods and siliceous poor grassland. Scattered.

Many petals

Least Snowbell

Soldanella minima
Primrose family (Primulaceae)
MAY–JULY 3–10CM ▼

Note Corolla white to pale violet, cut to one third (fringed).
Description Inside of flower darker. Stem with glandular hairs. Leaves rounded, less than 1cm diameter, rather thick and leathery.
Distribution Calcareous soils with long snow-cover, to 2500m. Rare.

Mountain Avens

Dryas octopetala
Rose family (Rosaceae)
MAY–AUG. 2–12CM ▼

Note Large, conspicuous flowers, with usually 8 petals.
Description Dwarf shrub with low growth habit. Flower stems hairy. Leaves leathery, with round-toothed margins, white-felted beneath.
Distribution Calcareous soils, grassland, scree and rocks. Scattered to common.

Stemless Carline Thistle

Carlina acaulis
Composite family (Asteraceae)
JUNE–SEPT. 3–20CM ▼

Note Large, almost sessile flowerhead surrounded by a ring of white bracts.
Description Flowerhead 5–12cm across. Stem short, with downy hairs. Leaves in rosette, pinnately-lobed and very spiny.
Distribution Nutrient-poor, dry soils to 2900m. Pastures and dry scrub. Scattered.

Many petals

Edelweiss

Leontopodium alpinum
Composite family (Asteraceae)
JULY–SEPT. 5–15CM ▼

Note Unmistakable. Star-shaped white woolly bracts surround flowerheads.
Description Flowerheads yellowish, 2–10 in tight central cluster. Leaves elongate, tongue-shaped, very hairy.
Distribution Rich calcareous soils to 3300m. Rock crevices and scree. Rare.

Carpathian Cat's-foot

Antennaria carpatica
Composite family (Asteraceae)
JUNE–AUG. 5–15CM

Note White flowerheads surrounded by brown, papery bracts.
Description Inflorescence a cluster of 2–6 small flowerheads. Leaves all narrowly-lanceolate, with white hairs.
Distribution Acid soils to 3200m, mainly in central ranges. Stony grassland, often on exposed sites. Scattered.

Yarrow

Achillea millefolium
Composite family (Asteraceae)
JUNE–OCT. 15–50CM

Note Leaves twice-pinnate and finely-divided.
Description Flowers white to pale pink. Flowerheads with 4–5 ray florets and yellowish disc florets. Stem ridged. Medicinal plant.
Distribution Rich soils. Meadows, from lowlands up to tree-line. Common.

Many petals

Broad-leaved Yarrow

Achillea macrophylla
Composite family (Asteraceae)
JULY–SEPT. 30–100CM ▼

Note Flowerheads to 1.5cm across. Bracts edged black.
Description Flowers in loose umbel-like panicles. Flowerheads with 5–8 ray florets. Leaves large, pinnate, large-lobed. Leaflets toothed.
Distribution Damp, nutrient-rich soils (northern slopes) to 2000m. Tall-herb communities and scrub. Scattered.

Silvery Yarrow

Achillea clavennae
Composite family (Asteraceae)
JUNE–SEPT. 5–30CM ▼

Note Whole plant covered with grey felty, appressed hairs.
Description Inflorescence a domed cluster of 5–20 flowerheads, each surrounded by 8–15 ray florets. Leaves deeply-cut.
Distribution Eastern Alps, on calcareous stony soils between 1500 and 2500m. Pastures, rock crevices and scree. Rare.

Black Yarrow

Achillea atrata
Composite family (Asteraceae)
JULY–SEPT. 5–30CM ▼

Note Narrow leaf lobes are finely-divided and pointed.
Description Plant faintly aromatic. Involucral bracts edged black. Outer ray florets 7–12.
Distribution Central and eastern Alps, west to Haute Savoie (Upper Savoy), to 2600m. Damp calcareous, stony grassland. Scattered.

Many petals

Musk Yarrow

Achillea moschata
Composite family (Asteraceae)
JULY–SEPT. 10–20CM ▼

Note Leaves divided comb-like.
Description Plant aromatic. Involucral bracts brownish. Outer ray florets 6–10.
Distribution Inner Alpine ranges, west to Mont Blanc, to 3200m. Damp, stony, acid soils. Scattered.

Dwarf Alpine Yarrow

Achillea nana
Composite family (Asteraceae)
JULY–SEPT. 5–15CM ▼

Note Plant strongly aromatic and densely hairy.
Description Involucral bracts dark brown, 6–8 outer ray florets. Leaves pinnate, glandular.
Distribution Western Alps, east to South Tyrol, 2000–3800m. Damp, stony grassland. Scattered.

False Aster

Aster bellidiastrum
Composite family (Asteraceae)
JUNE–SEPT. 5–30CM ▼

Note Resembles a large Daisy.
Description Flowerheads solitary. Rays often pink on underside. Stem leafless and hairy. Leaves oval, in rosettes.
Distribution Nutrient-rich, damp soils between 600 and 2800m. Pastures and stony scree slopes. Common.

Many petals

One-flowered Fleabane

Erigeron uniflorus
Composite family (Asteraceae)
JUNE–SEPT. 2–10CM

Note Involucral bracts upright, with woolly hairs.
Description Flowers solitary, white to pink. Upper stem hairy, lower stem glabrous. Leaves broadest towards tip.
Distribution Acid soils to 3500m. Ridges, rock crevices, rocky debris. Scattered.

Corymbose Tansy

Tanacetum corymbosum
Composite family (Asteraceae)
JUNE–AUG. 50–120CM ▼

Note Flowerheads corymbose (a raceme of flowerheads with pedicels of differing length so that all the flowerheads are at roughly the same level).
Description Flowerheads 2.5–5cm diameter. Leaves with 7–15 lobed and pointed segments.
Distribution Calcareous dry soils to 2000m. Meadows, open woodland and scrub. Scattered.

Alpine Moon-daisy

Leucanthemopsis alpina
Composite family (Asteraceae)
JUNE–AUG. 5–15CM

Note Daisy-like flowerheads to 4cm across.
Description Leaves broadly-oval in outline, pinnately-lobed. Stem leaves undivided.
Distribution Acid damp soils between 1600 and 3200m, mainly in central ranges. Snow-patches and scree. Scattered.

Many petals

Mountain Marguerite

Leucanthemum adustum
Composite family (Asteraceae)
JUNE–AUG. 15–30CM

Note At 4–7cm across, the flowerheads are noticeably bigger than those of Alpine Moon-daisy.
Description Leaves narrowly-oval in outline with 6–20 teeth on each side.
Distribution Mainly western Alps, east to a line from Lake Garda–Berchtesgaden, to 2300m. Scattered.

White Asphodel

Asphodelus albus
Lily family (Liliaceae)
MAY–JULY 60–120CM ▼

Note Dense, many-flowered spike-like raceme to 50cm tall.
Description Petals of equal length, to 2.5cm long. Flowering stem unbranched, leafless. Leaves grass-like, to 2cm broad, triangular and keeled.
Distribution Mainly on calcareous soils in southern Alps to over 2500m. Rare.

Scottish Asphodel

Tofieldia pusilla
Lily family (Liliaceae)
JULY–AUG. 5–12CM ▼

Note Small, rounded clusters of flowers, to 1.5cm across.
Description Flowers creamy-white. Leaves only at base of stem. Leaves grass-like, often growing close together.
Distribution Wet sandy or peaty soils between 1800 and 2400m. Spring-fed flushes and snow-patches. Rare.

Many petals

White False Helleborine

Veratrum album
Lily family (Liliaceae)
JUNE–AUG. 50–150CM ▼ ☠

Note Differs from Great Yellow Gentian in having alternate leaves.
Description Flowers to 2cm long, white inside, yellowish-green outside, in dense terminal spikes. Leaves oval, furrowed.
Distribution Damp nutrient-rich soils to over 2600m. Pastures and tall-herb communities. Scattered.

Lily-of-the-valley

Convallaria majalis
Lily family (Liliaceae)
MAY–JUNE 10–20CM ▼ ☠

Note Up to 10 bell-shaped drooping flowers in a one-sided raceme.
Description Flowers to 7mm, very fragrant. Pair of broadly-lanceolate leaves partially sheathing the stem. Berries red.
Distribution Dry deciduous woods and rocky debris between 600 and 2200m. Scattered, also naturalized.

Whorled Solomon's-seal

Polygonatum verticillatum
Lily family (Liliaceae)
JUNE–JULY 30–80CM ▼ ☠

Note Whorls of narrow leaves regularly spaced up the stem.
Description Flowers in clusters of 2–5, whitish with green teeth, arising from leaf axils.
Distribution Shady deciduous woods and tall-herb communities on damp soils between 800 and 1800m. Scattered.

Many petals

Twisted-stalk

Streptopus amplexifolius
Lily family (Liliaceae)
JUNE–JULY 30–100CM ▼

Note Zigzag stem with two rows of hairs.
Description Flowers to 1cm long, dangling beneath the leaves, tepal tips reflexed. Leaves oval, bases clasping stem. Berries red.
Distribution Shady damp woods and scrub, tall-herb communities, and forest clearings, to 2000m. Scattered.

St Bernard's Lily

Anthericum liliago
Lily family (Liliaceae)
MAY–JUNE 30–60CM

Note Inflorescence unbranched, and not one-sided.
Description Flowers to 22mm long. Inflorescence 6–10-flowered. Tepals of equal length. Leaves grass-like.
Distribution Western Alps to 1800m, east to Tyrol. Open woods, stony slopes, dry grassland. Scattered.

St Bruno's Lily

Paradisea liliastrum
Lily family (Liliaceae)
JUNE–JULY 30–50CM ▼

Note Funnel-shaped flowers droop to one side.
Description Perianth segments of unequal length. Flowers to 5cm long. Stem unbranched, leafless. Leaves grass-like, to 8mm wide.
Distribution Nutrient-rich calcareous soils to 2500m. Meadows, tall-herb communities and scrub. Scattered.

Many petals

Snowdon Lily

Lloydia serotina
Lily family (Liliaceae)
JUNE–AUG. 5–20CM ▼

Note Very dainty plant, with short stem leaves and solitary bell-shaped flowers.
Description Tepals with brownish-red stripes. Two grass-like basal leaves, similar in length to flower stalk.
Distribution Acid soils to 3000m, commonest in central Alps. Scattered.

Alpine Leek

Allium victorialis
Lily family (Liliaceae)
JUNE–AUG. 30–50CM ▼

Note Flowerheads globular, to 5cm across.
Description Flowers pale yellow-green. Flowerheads at first nodding. Stem rounded. Leaves 2–5cm broad.
Distribution Nutrient-rich, stony and base-poor soils between 1000 and 2500m. Meadows and tall-herb communities. Rare.

Pheasant's-eye

Narcissus poeticus ssp. *radiiflorus*
Lily family (Liliaceae)
MARCH–MAY 20–30CM ▼ ☠

Note Small, cup-shaped red-rimmed orange corona in centre of flower. Also known as Poet's Narcissus.
Description Tepals free or only slightly overlapping. Stem flattened. Leaves grass-like, grey-green.
Distribution Damp, nutrient-rich meadows in southern Alps to 2000m. Rare, but often grows in swarms. Also naturalized.

Zygomorphic flowers

Pale Clover

Trifolium pallescens
Legume family (Fabaceae)
JULY–AUG. 5–15CM ▼

Note Flowers often look a rather dirty white, pinkish towards base, and curving downwards after flowering.
Description Stem strong, to 3–10cm long, glabrous, often reddish.
Distribution Mainly in central and southern Alps, 1800–3000m on base-poor soils. Moraines and Alpine grassland. Common.

Mountain Clover

Trifolium montanum
Legume family (Fabaceae)
MAY–OCT. 15–50CM

Note Flowerheads consist of many small white flowers.
Description Flowerheads 1–1.5cm diameter on short 1.5mm stalks. Leaflets narrow, toothed and pointed; softly-hairy beneath.
Distribution Dry calcareous soils. Open woods and meadows between 400 and 2400m. Scattered.

Snow Clover

Trifolium pratense ssp. *nivale*
Legume family (Fabaceae)
JUNE–SEPT. 5–25CM

Note Flowerheads to 3.5cm, surrounded by upper stem leaves. This is a mountain subspecies of Red Clover, the latter common in lowlands.
Description Flowers white, with yellowish or pinkish tinge. Calyx very hairy. Leaflets oval.
Distribution Meadows and pastures between 1600 and 2800m. Common.

Zygomorphic flowers

Thal's Clover

Trifolium thalii
Legume family (Fabaceae)
JULY–AUG. 5–15CM

Note Flowers at first white, becoming reddish with age.
Description Plant not creeping. Stem only to 4cm long. Leaves glabrous, with small teeth. The common White Clover (Trifolium repens) is very similar, but has a creeping habit.
Distribution Stony, mainly calcareous soils between 1400 and 2800m. Meadows and pastures. Scattered.

Alpine Milk-vetch

Astragalus alpinus
Legume family (Fabaceae)
JUNE–AUG. 5–20CM

Note Flower wings white, keel and standard violet-tipped.
Description Leaves with 13–25 leaflets. Keel as long as standard. Wings entire. Pod hairy.
Distribution Nutrient-poor, calcareous soils in meadows and pastures, to over 3000m. Scattered.

Southern Milk-vetch

Astragalus australis
Legume family (Fabaceae)
JULY–AUG. 10–20CM

Note Flowers yellowish-white, violet only at tip of keel.
Description Leaves with 9–15 leaflets. Keel longer than standard. Wings two-lobed. Pod glabrous.
Distribution Dry, calcareous, stony soils between 1200 and 1800m. Poor grassland, rocky debris. Scattered.

Zygomorphic flowers

Yellow Alpine Milk-vetch

Astragalus frigidus
Legume family (Fabaceae)
JULY–AUG. 10–40CM

Note Flowers a uniform yellowish-white.
Description Calyx with short black hairs. Leaves with 9–17 blue-green leaflets. Stipules oval, to 1cm long. Pod with short hairs.
Distribution Dry, calcareous, stony soils, 1500–2700m. Steep slopes and grassland. Scattered.

Mountain Tragacanth

Astragalus sempervirens
Legume family (Fabaceae)
JULY–AUG. 5–20CM

Note Plant densely-thorny.
Description Flowers whitish-pink. Leaves with thorny tips, pinnate with 12–20 long hairy leaflets. Pod with short hairs.
Distribution Western Alps, east to Bernese Alps and Tessin. Calciphile, to 2700m. Broken grassland and rocky debris. Rare.

White Dead-nettle

Lamium album
Labiate family (Lamiaceae)
APRIL–OCT. 30–60CM

Note Leaves nettle-like, soft and without stinging hairs.
Description Flowers in whorls in the upper leaf axils. Upper lip spoon-shaped; lower lip two-lobed. Leaves opposite, toothed.
Distribution Nutrient-rich soils to 2000m. Meadows and soils enriched by livestock dung. Common.

Zygomorphic flowers

Mountain Germander

Teucrium montanum
Labiate family (Lamiaceae)
JUNE–SEPT. 10–30CM

Note White flowers, lacking upper lip.
Description Lower stem woody. Leaves grey-green, with obvious central vein, rolled margins, and softly-hairy beneath.
Distribution Throughout the Alps, to 2400m on calcareous soils. Dry grassland and rocky slopes. Scattered.

Common Eyebright

Euphrasia rostkoviana
Figwort family (Scrophulariaceae)
JUNE–OCT. 5–25CM

Note Flowers white-violet, with yellow spot in the throat.
Description Plant branching from lower half. Upper lip of flower with violet markings. Leaves oval, unstalked, with 9–13 teeth.
Distribution Meadows and pastures between 400 and 2600m. Scattered.

Alpine Butterwort

Pinguicula alpina
Butterwort family (Lentibulariaceae)
MAY–AUG. 5–20CM ▼

Note Catches insects and other invertebrates on its sticky, fleshy leaves.
Description Flowers white, with 1–3 yellow patches on the lower lip. Spur 2–3mm long. Leaves yellow-green, with slightly upturned margins.
Distribution Nutrient-poor, calcareous soils. Wet rocks, mires and grassland, to 2500m. Scattered.

Zygomorphic flowers

Marsh Helleborine

Epipactis palustris
Orchid family (Orchidaceae)
JUNE–AUG. 30–50CM ▼

Note Inner petals white, sepals brownish-green.
Description Flowers tend to face the same way. Lip with frilly margin. Leaves lanceolate.
Distribution Damp, calcareous, humus-rich soils. Damp ground and fens to 2000m. Rare.

Lesser Butterfly-orchid

Platanthera bifolia
Orchid family (Orchidaceae)
MAY–AUG. 15–50CM ▼

Note Narrow spur, up to 2cm long, is characteristic.
Description Flowers sweetly fragrant, especially in the evening. Lip tongue-shaped, angled downwards. Paired basal leaves elongate-oval.
Distribution Nutrient-poor, damp soils subject to periodic drying, to 2100m. Open woods and poor grassland. Scattered.

Creeping Lady's-tresses

Goodyera repens
Orchid family (Orchidaceae)
JULY–AUG. 10–30CM ▼

Note Spreads by stolons, forming colonies.
Description Spike many-flowered. Flowers only 3–6mm long. Leaves in rosette, net-like veins on upper side.
Distribution Moss-rich, fairly dry to damp coniferous woods, to 2000m. Very rare.

Four petals

Alpine Clematis

Clematis alpina
Buttercup family (Ranunculaceae)
MAY–JULY 100–250CM ▼

Note Stalks twine with tendrils.
Description Flowers long-stalked, 4–6mm diameter, with 10–20 petaloid stamens in the centre. Leaves 2-ternate.
Distribution Mainly on calcareous substrates. In partial shade amongst rose bushes or Mountain Pine scrub. Scattered.

Bluish Rock-cress

Arabis caerulea
Crucifer family (Brassicaceae)
JUNE–AUG. 5–12CM

Note Leaves spatulate, with a few, rather obvious teeth.
Description Flowers pale violet, 5mm long, 2–8 in a raceme. Pod 1–3cm.
Distribution In Alps between 2000 and 3300m. Damp, calcareous soils. Snow-patches, scree and rocky debris. Scattered to rare.

Five-leaved Coral-wort

Cardamine pentaphyllos
Crucifer family (Brassicaceae)
APRIL–JUNE 20–50CM

Note Lower stem-leaves palmate.
Description Flowers blue-violet, to 2cm long, in a terminal raceme. Stem-leaves alternate, stalked, with very small bulbils in the axils.
Distribution Mainly in Beech woods in the eastern Alps, to 2000m. Scattered.

Four petals

Cross Gentian

Gentiana cruciata
Gentian family (Gentianaceae)
JULY–AUG. 20–70CM ▼

Note Leaves in opposite pairs, close together on lower stem.
Description Flowers bell-shaped, to 2.5cm long, 4-lobed. Leaves opposite, leathery.
Distribution Mainly in calcareous Alps to 1600m. Dry grassland, scrub and woodland edges. Rare.

Fringed Gentian

Gentianella ciliata
Gentian family (Gentianaceae)
AUG.–OCT. 10–25CM ▼

Note Flower lobes fringed with fine hairs.
Description Flowers mostly solitary, 2–5cm long. Corolla lobes somewhat inrolled. Calyx tips pointed. Leaves narrowly-lanceolate, lacks basal leaves.
Distribution Almost restricted to calcareous Alps to 2200m. Poor grassland and pastures. Scattered.

Slender Gentian

Gentianella tenella
Gentian family (Gentianaceae)
JULY–SEPT. 2–10CM ▼

Note Flowers pale blue, bearded in centre, to 1cm long.
Description Corolla lobes pointed, upright. Basal leaves usually withered by flowering time.
Distribution Nutrient-rich, acid soils between 1800 and 3000m. Patchy grassland, soils enriched by livestock dung, and rocky debris. Rare.

Four petals

Dwarf Gentian

Gentianella nana
Gentian family (Gentianaceae)
JULY–SEPT. 2–7CM ▼

Note Very small, flowers to 7mm long, bearded at centre.
Description Corolla spreading, 4–5-lobed. Lobes rounded and slightly pointed.
Distribution Only in eastern central Alps from Tyrol to Carinthia, on damp, acid, stony soils, from 2200 to 2800m. Rare.

Germander Speedwell

Veronica chamaedrys
Figwort family (Scrophulariaceae)
APRIL–JULY 15–30CM

Note Easy to recognize by the two opposite lines of hairs along the stem.
Description Racemes of flowers in the upper leaf axils, each with 10–20 flowers. Flowers about 12mm across.
Distribution Nutrient-rich soils to 2200m. Meadows, pastures and woodland edges. Scattered. Common in lowland.

Rock Speedwell

Veronica fruticans
Figwort family (Scrophulariaceae)
JUNE–AUG. 5–15CM

Note Centre of flower has a white spot, ringed red.
Description Spreading habit; stems woody at base. Flowers deep blue. Calyx 5-lobed. Leaves oval, rather thick and shiny.
Distribution Dry stony soils. Rocky debris, scree and loose grassland, to 2800m. Scattered.

Four petals

Alpine Speedwell

Veronica alpina
Figwort family (Scrophulariaceae)
JUNE–AUG. 2–15CM

Note Inflorescence umbel-like, with 5–20 flowers.
Description Flowers 5–8mm diameter, calyx usually 5-lobed. Leaves oval. Plant hairy or glabrous.
Distribution Damp, nutrient-rich soils to over 3000m. Pastures and tall-herb communities. Common, but rare in southern Alps.

Leafless-stemmed Speedwell

Veronica aphylla
Figwort family (Scrophulariaceae)
JUNE–AUG. 3–7CM

Note Stem leafless except for small flower bracts.
Description Flowers 1–6 per stem, 6–8mm diameter. Calyx 4-lobed with glandular hairs. Leaves densely-hairy.
Distribution Mainly in calcareous Alps to 2800m. Stony grassland and snow-patches. Scattered.

Violet Speedwell

Veronica bellidioides
Figwort family (Scrophulariaceae)
JUNE–AUG. 5–20CM

Note Basal leaves in rosette.
Description Inflorescence umbel-like, with 3–10 violet-blue flowers. Calyx 4-lobed. Rosette leaves to 4cm long, stem leaves narrower and shorter.
Distribution Nutrient- and lime-poor grassland to 3200m. Common in central Alps, absent from calcareous Alps.

Four petals

Heath Speedwell

Veronica officinalis
Figwort family (Scrophulariaceae)
JUNE–AUG. 15–30CM

Note Flowers pale lilac with darker veins.
Description Flowers in racemes arising from leaf axils. Stems creeping to ascending. Leaves short-stalked, toothed and softly hairy.
Distribution Nutrient-poor, dry soils to 2000m. Dry woods, scrub and grassland. Scattered.

Thyme-leaved Speedwell

Veronica serpyllifolia
Figwort family (Scrophulariaceae)
JUNE–AUG. 5–15CM

Note Flowers pale blue with darker veins.
Description Flowers 5–6mm diameter, 5–20 in loose racemes arising from leaf axils. Leaves oval, entire to finely-toothed.
Distribution Damp, nutrient-rich soils to 2000m. Spring-fed flushes and soils enriched by livestock dung. Scattered.

Compact Blue Speedwell

Veronica allionii
Figwort family (Scrophulariaceae)
JUNE–AUG. 5–15CM

Note Flowers in dense spikes, rounded at the top.
Description Flowers with short corolla tube. Stem creeping, woody at base. Leaves glabrous and leathery.
Distribution Only in south-western Alps, to 2500m. Pastures and open woods. Scattered.

Four petals

Spiked Speedwell

Veronica spicata
Figwort family (Scrophulariaceae)
JULY–SEPT. 10–40CM ▼

Note Inflorescence a many-flowered, long, pointed, tapering raceme.
Description Flowers to 12mm diameter, with long corolla tube. Stem densely-hairy. Leaves lanceolate, mostly opposite.
Distribution Warm, nutrient-rich soils to 1500m. Grassland and scrub. Scattered.

Bluish Paederota

Paederota bonarota
Figwort family (Scrophulariaceae)
JULY–AUG. 10–20CM ▼

Note Flowers in dense, usually drooping, racemes.
Description Flowers tubular, 4-partite to 2-lipped, to 1.5cm long. Stem densely-hairy. Leaves with forward-pointing teeth.
Distribution Only in eastern Alps, notably in southern calcareous Alps to 2600m. Rock crevices. Rare.

Devils-bit Scabious

Succisa pratensis
Teasel family (Dipsacaceae)
JULY–SEPT. 30–100CM

Note Rounded, bluish-pale violet flowerheads with prominent stamens.
Description Flowerheads to 2.5cm diameter, surrounded by a ruff of pointed bracts.
Distribution Periodically damp sites, fens, open woodland and scrub, rarely above 1500m. Scattered.

Five petals

Alpine Columbine

Aquilegia alpina
Buttercup family (Ranunculaceae)
JUNE–AUG. 20–70CM ▼ ☠

Note Flowers pale blue with almost straight spurs.
Description Flowers 5–8cm diameter. Stem scarcely branched, 1–3-flowered. Basal leaves 2-ternate.
Distribution Western Alps, to 2500m. Mainly on damp, calcareous soils. Scrub and open woodland. Rare.

Common Columbine

Aquilegia vulgaris
Buttercup family (Ranunculaceae)
MAY–JULY 30–70CM ▼ ☠

Note Flowers blue-violet with hooked spurs.
Description Flowers to 5cm diameter, up to 10 per branching stem. Leaves 2-ternate.
Distribution Throughout the Alps, but nowhere common. Open woods, scrub, meadows, to 2000m.

Alpine Larkspur

Delphinium elatum
Buttercup family (Ranunculaceae)
JULY–AUG. 60–200CM ▼ ☠

Note Flowers blue, with backward- and upward-pointing spur.
Description Attractive plant, with flowers in racemes. Leaves large and palmately-divided. Very poisonous!
Distribution Only in central and southern ranges. Woodland edge, tall-herb communities and riverbanks. Very rare.

Five petals

Sticky Primrose

Primula glutinosa
Primrose family (Primulaceae)
JUNE–AUG. 2–8CM ▼

Note Easily identified by the sticky, jagged-toothed leaves.
Description Flowers strongly-scented. Flower colour ranges from blue to lilac, changing as the flowers age.
Distribution Central and southern eastern Alps, from eastern Switzerland to Carinthia, on lime-poor soils, to 3100m. Rare.

Alpine Flax

Linum alpinum
Flax family (Linaceae)
JUNE–JULY 10–30CM

Note Petals blue, white and yellow towards centre of flower.
Description Flowers to 2.5cm diameter. Flowering stems leafy, non-flowering stems less leafy.
Distribution Dry, stony soils on limestone. Rocky debris, scree and loose grassland, to 2800m. Scattered.

Alpine Eryngo

Eryngium alpinum
Umbellifer family (Apiaceae)
JULY–SEPT. 30–80CM ▼

Note Spiny plant with bluish sheen towards the top.
Description Flowers small and numerous, in large, cylindrical flowerheads.
Distribution Only in southern and western Alps. Calcareous, stony soils, to 2500m. Scrub, meadows and pastures. Scattered, also grown as an ornamental.

Five petals

Willow Gentian

Gentiana asclepiadea
Gentian family (Gentianaceae)
JUNE–SEPT. 30–70CM ▼

Note Flowers grow amongst leaves in the upper axils.
Description Flowers to 5cm long, with reddish-violet spots inside. Plant very leafy. Leaves opposite.
Distribution Damp, humus-rich, calcareous soils to over 2000m. Open woods, riverbanks. Scattered.

Stemless Gentian

Gentiana acaulis
Gentian family (Gentianaceae)
MAY–AUG. 5–10CM ▼

Note Flowers with olive green longitudinal stripes inside.
Description Corolla to 6cm long. Usually 1–2 pairs of stem leaves. Basal leaves soft, 4–10cm long.
Distribution Nutrient- and lime-poor soils between 1200 and 3000m. Pastures and Alpine grassland. Scattered in the central Alps, otherwise very rare.

Trumpet Gentian

Gentiana clusii
Gentian family (Gentianaceae)
MAY–AUG. 5–12CM ▼

Note Flowers with pale violet spots inside.
Description Corolla to 6cm long. Usually 1–2 pairs of stem leaves. Basal leaves hard, 3–5cm long.
Distribution Calcareous, damp and stony soils to 2800m. Poor grassland, rocky debris. Common in northern Alps, becoming rarer towards the south-west.

Five petals

Alpine Gentian

Gentiana alpina
Gentian family (Gentianaceae)
MAY–AUG. 4–8CM ▼

Note Flowers with olive green markings inside. Leaves 2cm long maximum.
Description Corolla to 4cm long. Usually lacking stem leaves. Basal leaves yellow-green, leathery and fleshy.
Distribution Stony, acid soils to 2600m. Poor grassland and rocks. Scattered in south-western Alps.

Snow Gentian

Gentiana nivalis
Gentian family (Gentianaceae)
JULY–SEPT. 5–20CM ▼

Note All stems produce flowers.
Description Stems mostly many-flowered, branching from base. Flowers to 2cm long. Upper leaves lanceolate and pointed, basal leaves small and oval.
Distribution Nutrient-poor, calcareous soils between 1600 and 2900m. Meadows and pastures. Scattered.

Bavarian Gentian

Gentiana bavarica
Gentian family (Gentianaceae)
JULY–SEPT. 5–20CM ▼

Note Leaves oval and broadest above the middle, entire and glabrous.
Description Flowers to 3cm long. Lower leaves tightly-packed but not in rosette.
Distribution Damp soils with long snow-cover, between 1600 and 3000m. Snow-patches, spring-fed flushes. Scattered.

Five petals

Short-leaved Gentian

Gentiana brachyphylla
Gentian family (Gentianaceae)
JULY–AUG. 3–7CM ▼

Note Leaves very small (<1cm long) grow in rosettes.
Description Flowers solitary, to 2cm long. Corolla tube narrow, lobes with greenish shimmer on outside.
Distribution Stony, acid, damp soils between 2000 and 3200m. Rock debris and patchy grassland. Rare.

Bladder Gentian

Gentiana utriculosa
Gentian family (Gentianaceae)
MAY–AUG. 10–20CM ▼

Note Calyx inflated and almost as long as corolla tube.
Description Flowers to 2.5cm long. Stem usually branching, many-flowered. Calyx with winged edges. Basal leaves mostly withered by flowering time.
Distribution Wet, calcareous, humus-rich soils. Spring-fed flushes and fens to 2500m. Rare.

Spring Gentian

Gentiana verna
Gentian family (Gentianaceae)
APRIL–AUG. 3–12CM ▼

Note Flowers earlier than other gentians.
Description Flowers to 3cm long. Stem unbranched, single-flowered, with 1–3 pairs of leaves. Calyx with winged edges.
Distribution Calcareous, nutrient-poor soils between 600 and 2900m. Meadows and pastures. Scattered.

Five petals

Carinthian Lomatogonium

Lomatogonium carinthiacum
Gentian family (Gentianaceae)
AUG.–SEPT. 3–15CM ▼

Note Pale blue star-shaped flowers, petals joined near base.
Description Flowers to 2cm diameter, solitary, on leafless, square stems. Leaves not in rosette, oval to lanceolate.
Distribution Stony, poor soils, from Valais to the lower Tauern, between 1400 and 2500m. Very rare.

Marsh Felwort

Swertia perennis
Gentian family (Gentianaceae)
JULY–SEPT. 15–50CM ▼

Note Petals open into star shape, with dark stripes and spots on upper surface.
Description Flowers in loose panicles. Stem square. Lower leaves opposite, upper leaves alternate.
Distribution Throughout Alps, but rare everywhere.

Jacob's Ladder

Polemonium caeruleum
Phlox family (Polemoniaceae)
JUNE–AUG. 30–120CM ▼

Note Leaves pinnate, ascending the stem in step-like pattern – hence the common name.
Description Corolla sky blue with white centre, to 2cm across, fused at base. Stem angled and hollow.
Distribution Nutrient-rich, damp soils to 2300m. Meadows, tall-herb communities and riverbanks. Rare.

Five petals

Viper's Bugloss

Echium vulgare
Borage family (Boraginaceae)
JUNE–SEPT. 20–100CM

Note Whole plant covered with slightly prickly, hard, bristly hairs.
Description Flowers pink in bud, opening to blue-violet or blue. Stem with dark spots. Leaves narrowly-lanceolate.
Distribution Dry, stony, nutrient-rich soils to 2200m. Meadows and footpaths. Common.

King of the Alps

Eritrichium nanum
Borage family (Boraginaceae)
JULY–AUG. 1–5CM ▼

Note Flowers resemble those of a forget-me-not, but this plant forms cushions.
Description Flower has yellow ring at the centre and bracts below. Leaves only to 1cm long, with silky hairs.
Distribution Mostly acid, stony soils, between 2000 and 3000m. Rock crevices and scree. Rare.

Alpine Forget-me-not

Myosotis alpestris
Borage family (Boraginaceae)
MAY–SEPT. 5–25CM

Note Young pinkish-violet flowers and older blue flowers often together in the same inflorescence.
Description Flowers without bracts. Yellow ring at centre of flower. Leaves stalked, to over 2cm long.
Distribution Nutrient-rich, usually somewhat moist soils. Stony grassland, pastures and herb communities. Common.

Five petals

South-alpine Lungwort

Pulmonaria australis
Borage family (Boraginaceae)
APRIL–JUNE 15–40CM ▼

Note Upper surface of leaves with hairs of different length.
Description Flowers pink at first, later blue. Corolla five-lobed, divided at most to middle. Leaves unspotted.
Distribution Central and southern Alps to 2600m. Open woods, grassland and dwarf-shrub heath. Very rare.

Fairy Foxglove

Erinus alpinus
Figwort family (Scrophulariaceae)
APRIL–JUNE 5–20CM ▼

Note Two of the five conjoined petals are somewhat narrower and shorter.
Description Leaves with glandular hairs. Basal leaves in rosette, stem leaves alternate.
Distribution Calcareous soils, 600–2300m. Rock crevices and scree. Scattered.

Carinthian Wulfenia

Wulfenia carinthiaca
Figwort family (Scrophulariaceae)
JULY–AUG. 20–50CM ▼

Note Inflorescence a dense panicle growing from a large basal rosette.
Description Panicle bends to one side. Stem with small, inconspicuous leaves. Rosette leaves to 15cm long, obovate.
Distribution Only on the border between Carinthia (Austria) and Friuli (Italy). Locally not rare.

Five petals

Round-leaved Rampion

Phyteuma globulariifolium
Bellflower family (Campanulaceae)
JULY–SEPT. 3–7CM

Note Dainty plant with flowerheads of 2–7 strongly-curved flowers each to 2cm long.
Description Leaves narrowly-oval, only 2–4 times as long as broad, untoothed or toothed only at tip.
Distribution Stony, acid, humus-rich soils to over 3000m. Ridges, rock crevices and scree. Very rare.

Globe-headed Rampion

Phyteuma hemisphaericum
Bellflower family (Campanulaceae)
JULY–AUG. 5–30CM

Note Leaves grass-like and grooved.
Description Flowers to 1.5cm long, claw-like in hemispherical inflorescence. Leaves entire.
Distribution Mainly central Alpine ranges between 1700 and 3000m. Rock crevices and stony grassland. Common on dry, acid soils.

Round-headed Rampion

Phyteuma orbiculare
Bellflower family (Campanulaceae)
MAY–SEPT. 10–40CM

Note Flowers to 1.5cm long, 15–30 in rounded flowerheads.
Description Basal leaves stalked, oval to narrowly-lanceolate, toothed. Upper stem leaves unstalked.
Distribution Lime-rich, stony soils to 2500m. Poor grassland, fens and open woods. Common.

Five petals

Dolomite Rampion

Phyteuma sieberi
Bellflower family (Campanulaceae)
JULY–AUG. 5–30CM

Note Flowers 5–15 in rounded flowerhead.
Description Basal leaves round-oval to broadly-lanceolate, hairy. Upper stem leaves unstalked.
Distribution Only in southern calcareous Alps between 1600 and 2600m. Rocks and stony grassland. Scattered.

Horned Rampion

Phyteuma scheuchzeri
Bellflower family (Campanulaceae)
MAY–JULY 10–40CM

Note Rounded flowerheads on long stalks.
Description Flowers only slightly curved. Involucral bracts long and prominent. Stem leaves very narrow, with toothed margins.
Distribution Rock crevices in central and southern Alps between 600 and 2200m. Scattered.

Dark Rampion

Phyteuma ovatum
Bellflower family (Campanulaceae)
JUNE–JULY 30–80CM

Note Flowers very dark violet.
Description Flowerheads oval, becoming cylindrical. Flowers with 2 stigmas. Upper stem leafy. Basal leaves long-stalked, heart-shaped and toothed.
Distribution Damp, nutrient-rich soils between 1200 and 2100m. Meadows and tall-herb communities. Scattered to rare.

Five petals

Blue-spiked Rampion

Phyteuma betonicifolium
Bellflower family (Campanulaceae)
JULY–AUG. 20–60CM

Note Flowers blue, upright before opening, in an oval to cylindrical spike.
Description Basal leaves stalked, narrowly heart-shaped. Stem leaves narrow, the upper leaves very small and unstalked.
Distribution Lime- and nutrient-poor soils to 2600m. Poor grassland, woodland edges. Scattered to common.

Spiked Bellflower

Campanula spicata
Bellflower family (Campanulaceae)
JUNE–AUG. 15–80CM ▼

Note Unmistakable multi-flowered, long, loose spike.
Description Stem unbranched. Plant with rough hairs. Leaves narrowly-lanceolate, unstalked.
Distribution Warm, rocky, calcareous soils. Poor grassland, rocks and rocky debris. Scattered.

Bearded Bellflower

Campanula barbata
Bellflower family (Campanulaceae)
JUNE–AUG. 10–50CM

Note Insides of flowers with long hairs.
Description Flowers, to 3cm long, nodding to one side. Stem unbranched. Leaves narrowly-oval, with bristly hairs.
Distribution Nutrient- and lime-poor soils between 1200 and 2700m. Pastures, grassland and scrub. Scattered.

Five petals

Clustered Bellflower

Campanula glomerata
Bellflower family (Campanulaceae)
MAY–SEPT. 15–70CM

Note Violet flowers clustered at the ends of stems and in the axils of the upper stem leaves.
Description Style shorter than corolla. Lower stem leaves stalked; upper almost unstalked.
Distribution Nutrient-rich, calcareous, loamy soils to 1800m. Scattered.

Nettle-leaved Bellflower

Campanula trachelium
Bellflower family (Campanulaceae)
JULY–AUG. 50–100CM

Note Leaves resemble those of Common Nettle.
Description Flowers to 5cm, with ciliate hairs around tips. Stem angled. Leaves and stem with stiff hairs. Lower stem leaves stalked, upper leaves unstalked.
Distribution Throughout Alps to 1700m. Nutrient-rich soils in woods and scrub. Scattered.

Alpine Bellflower

Campanula alpina
Bellflower family (Campanulaceae)
JUNE–AUG. 5–20CM ▼

Note Flowers down to very low on the stem.
Description Corolla hairy at tips. Leaves and stem with white-woolly hairs. Leaves linear, with white midrib.
Distribution Only in eastern Alps, to 2400m, mainly on lime-poor soils. Rare.

Five petals

Zois' Bellflower

Favratia zoysii
Bellflower family (Campanulaceae)
JULY–AUG. 2–10CM ▼

Note Flowers tapering towards tip, and almost closed.
Description Flowers to 2cm long, with white hairs within. Basal leaves small, spoon-shaped. Plant glabrous.
Distribution Only in south-eastern Alps between 1500 and 2300m, on limestone. Rock crevices and debris. Rare.

Mount Cenis Bellflower

Campanula cenisia
Bellflower family (Campanulaceae)
JULY–SEPT. 2–5CM ▼

Note Flowers funnel-shaped, with wide-opening corolla lobes.
Description Corolla cut to halfway. Stem leafy up to flowers. Leaves oval, entire.
Distribution Western Alps, to eastern Tyrol, to over 3000m. Limestone debris and moraines. Rare.

Fairy's Thimble

Campanula cochleariifolia
Bellflower family (Campanulaceae)
JUNE–SEPT. 5–15CM ▼

Note Rounded, almost spoon-shaped basal leaves and linear stem leaves.
Description Flowers only slightly cut. Stem leaves serrate-toothed with short, bristly hairs.
Distribution Stony, calcareous, seeping-damp soils. Rock crevices and debris. Scattered.

Five petals

Broad-leaved Harebell

Campanula rhomboidalis
Bellflower family (Campanulaceae)
JULY–AUG. 20–70CM

Note Calyx lobes linear and spreading beneath open bell-shaped flowers.
Description Stem angled, glabrous, or only hairy on ridges. Leaves elongate-lanceolate.
Distribution Western Alps, east to the Rhine, to 1800m. Damp meadows on calcareous soils. Rare.

Scheuchzer's Bellflower

Campanula scheuchzeri
Bellflower family (Campanulaceae)
JULY–AUG. 5–40CM

Note The large bell-shaped flowers look out of proportion for such a delicate plant.
Description Flowers nodding before fully open. Stem leaves unstalked, narrowly-lanceolate, with ciliate hairs at the base.
Distribution Nutrient- and lime-poor, stony soils to 3100m. Grassland, rocky debris and heath. Scattered.

Grass-like Bellflower

Campanula cespitosa
Bellflower family (Campanulaceae)
JUNE–AUG. 15–70CM

Note Narrow, elongate leaves clustered around lower stem.
Description Calyx lobes not reflexed. Rosette leaves rounded, narrowing towards stalk.
Distribution Eastern Alps on limestone, west to Brenner Pass, to 3000m. Rock crevices and rocky debris. Scattered.

Many petals

Hepatica

Hepatica nobilis
Buttercup family (Ranunculaceae)
APRIL–JUNE 5–15CM ▼ (☠)

Note Calyx-like bracts below flowers.
Description Flowers with 6–10 petals. Leaves three-lobed, reddish-brown beneath, appearing after flowering.
Distribution Calcareous soils in deciduous woods or scrub, from lowland to 1900m. Scattered.

Alpine Aster

Aster alpinus
Composite family (Asteraceae)
JUNE–SEPT. 5–30CM ▼

Note Flowerheads conspicuous, bluish- to pinkish-violet, to 5cm across, solitary on hairy stems.
Description Central disc florets yellow, surrounded by long ray florets. Leaves entire, hairy.
Distribution Nutrient-poor, mainly calcareous soils, 900 to 3000m. Poor grassland and rocks. Scattered.

Pygmy Gentian

Gentiana prostrata
Gentian family (Gentianaceae)
JUNE–AUG. 2–8CM ▼

Note Pale blue flowers divided into 8–10 lobes.
Description Calyx 5-lobed, but not winged. Stems mainly lie along the ground. Leaves oval, growing over each other.
Distribution Damp, nutrient-rich, stony soils to 2800m. Grassland and rocky debris. Rare.

Many petals

Perennial Cornflower

Centaurea montana
Composite family (Asteraceae)
MAY–OCT. 20–60CM ▼

Note Flower 2-coloured – outer florets blue, inner florets red.
Description Flower bracts with blackish-brown fringe around edge. Leaves lanceolate, finely-toothed, growing close to stem.
Distribution Damp, nutrient-rich soils, especially on limestone, to 2300m. Alpine grassland, pastures and scrub. Scattered to common.

Squarrose Knapweed

Centaurea triumfettii
Composite family (Asteraceae)
JUNE–AUG. 15–60CM ▼

Note Leaves with grey, felty hairs on both sides.
Description Flower bracts with long, brownish fringe. Leaves narrowly-lanceolate, with grey felty hairs on both sides. Leaf stalks short.
Distribution Dry, calcareous soils to 2000m. Mountain meadows, open woodland and scrub. Scattered.

Alpine Saw-wort

Saussurea alpina
Composite family (Asteraceae)
JULY–AUG. 5–20CM ▼

Note Flowerheads 5–10, in compact, umbel-like cluster.
Description Flowerhead 1cm diameter. Leaves lanceolate and entire, felty beneath, but not white.
Distribution Stony, mainly lime-poor soils between 1800 and over 3000m. Ridges, rock crevices and rocky debris. Rare.

Many petals

Heart-leaved Saussurea

Saussurea discolor
Composite family (Asteraceae)
JULY–SEPT. 5–30CM ▼

Note Undersides of leaves snow-white.
Description Flowers in heads of 3–8, strongly vanilla-scented. Stem with felty hairs. Leaves mostly narrow, triangular, pointed, and irregularly-toothed; lower leaves heart-shaped at base.
Distribution Stony, calcareous soils to 2800m. Stony grassland and rock crevices. Rare.

Dwarf Saussurea

Saussurea pygmaea
Composite family (Asteraceae)
JULY–AUG. 5–20CM ▼

Note Stem felty, with terminal flowerhead.
Description Flowerhead about 3cm diameter. Stem felty-hairy. Leaves linear-lanceolate, rough, grey beneath.
Distribution Only in eastern Alps, to 2500m, on calcareous soils. Stony grassland, rock crevices and rocky debris. Rare.

Alpine Sow-thistle

Cicerbita alpina
Composite family (Asteraceae)
JULY–SEPT. 50–200CM

Note Flowers in a terminal panicle on reddish-brown, densely hairy stem.
Description Leaves pinnately-lobed, with a large, triangular terminal lobe; blue-green beneath.
Distribution Damp, nutrient-rich soils, especially on limestone. Between 1000 and 2000m. Tall-herb communities and woods. Scattered.

Zygomorphic flowers

Monk's-hood

Aconitum napellus
Buttercup family (Ranunculaceae)
JUNE–AUG. 30–200CM ▼ ☠

Note Helmet of flower rounded, as broad or broader than high.
Description Flowers in raceme, occasionally branched. Flowers uniformly bluish-violet, rarely white. Leaves divided almost to the midrib. Very poisonous.
Distribution Throughout the Alps to 2500m. Tall-herb communities, soils enriched by dung, riverbanks. Common.

Variegated Monk's-hood

Aconitum variegatum
Buttercup family (Ranunculaceae)
JULY–SEPT. 30–200CM ▼ ☠

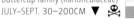

Note Helmet of flower higher than broad.
Description Inflorescence usually branched. Flowers pale violet, often streaked white. Leaves with network of veins. Very poisonous.
Distribution More easterly, to 2200m. Damp, nutrient-rich soils. Tall-herb communities, riverbanks. Scattered.

Long-spurred Pansy

Viola calcarata
Violet family (Violaceae)
JUNE–AUG. 3–12CM ▼

Note Flower with upwardly arching spur, 8–15mm long.
Description Flowers 2.5–4cm, with a dark striped yellow patch at the centre. Leaves oval, all basal, with 1–3 notches at each side.
Distribution Western Alps, east to Tyrol. Stony soils to 3000m. Pastures and rocky debris. Common.

Zygomorphic flowers

Alpine Pansy

Viola alpina
Violet family (Violaceae)
MAY–JULY 4–10CM

Note Flowers violet, with a white patch inside and dark ray-like stripes.
Description Flowers to 3cm, with a 3–4mm short spur. Leaves oval, all basal.
Distribution Only in north-eastern calcareous Alps, to 2200m. Stony soils. Poor grassland and rocky debris. Scattered to rare.

Mount Cenis Pansy

Viola cenisia
Violet family (Violaceae)
JUNE–AUG. 5–20CM ▼

Note Leaves entire, not notched.
Description Plant spreading. Flowers 2–2.5cm diameter, unstriped, with small yellow spot at the centre. Spur 5–10mm long. Leaves broadly-oval.
Distribution Only in western Alps, to 2900m. Mainly in rocky, calcareous sites. Rare.

Finger-leaved Violet

Viola pinnata
Violet family (Violaceae)
MAY–JUNE 3–10CM ▼

Note Leaves deeply palmately-lobed, with 5–9 lobes.
Description Flowers 1–1.5cm long, fragrant, pale violet, with hairy throat. Leaves basal and long-stalked.
Distribution Between 1000 and 2500m. Nutrient-poor, damp, calcareous soils. Poor grassland, rock crevices and rocky debris. Rare.

Zygomorphic flowers

Marsh Violet

Viola palustris
Violet family (Violaceae)
MAY–JULY 5–15CM

Note Flowers pale violet, with dark veins spreading out from the centre.
Description Flowers to 2cm long. Leaves basal, rounded, with wavy margin.
Distribution To 1800m, more rarely to 2500m. Wet, acid, often peaty soils. Spring-fed flushes, bogs, ditches and banks. Scattered.

Tufted Vetch

Vicia cracca
Legume family (Fabaceae)
JUNE–AUG. 30–150CM

Note Flowers bluish-violet, 10–40 in one-sided raceme.
Description Inflorescence long-stalked. Leaves with 12–30 leaflets and a terminal tendril.
Distribution Nutrient-poor, loamy soils to 2000m. Meadows and scrub. Scattered.

Mountain Milk-vetch

Oxytropis jacquinii
Legume family (Fabaceae)
JULY–AUG. 5–15CM

Note Keel of flower distinctly pointed.
Description Leaves stalked, with 15–35 oval-lanceolate silky-hairy leaflets. Fruit inflated.
Distribution Stony, calcareous soils between 1500 and 2800m. Grassland, ridges and rocky debris. Scattered.

Zygomorphic flowers

Purple Oxytropis

Oxytropis halleri
Legume family (Fabaceae)
MAY–JULY 5–20CM

Note Plant covered in long, erect hairs.
Description Stemless. Flowers lilac to pale violet. Leaves unstalked.
Distribution Dry, often exposed sites between 1500 and 2700m. Ridges and open grassland. Rare.

Bitter Milkwort

Polygala amara
Milkwort family (Polygalaceae)
MAY–AUG. 5–20CM

Note Oval rosette leaves are larger than the pointed stem leaves.
Description Flowers bluish-lilac with white hairy fringe. Plant prostrate, branching at base. Leaves taste bitter.
Distribution Calcareous soils to 2200m. Stony grassland, open woods and spring-fed flushes. Scattered.

Alpine Milkwort

Polygala alpestris
Milkwort family (Polygalaceae)
JUNE–JULY 5–15CM

Note Stem leaves increasingly long up the stem.
Description Flowers blue with white hairy fringe. Stem leaves only, lower leaves close together; no rosette.
Distribution Dry, calcareous soils between 1200 and 2700m. Poor grassland and heath. Scattered.

Zygomorphic flowers

Pyramidal Bugle

Ajuga pyramidalis
Labiate family (Lamiaceae)
APRIL–JUNE 5–30CM

Note Upper leaf-like bracts mainly suffused violet and longer than the flowers.
Description Leaf rosette with leaves larger than the stem leaves.
Distribution Nutrient- and lime-poor Alpine grassland between 1300 and 2700m. Scattered.

Bugle

Ajuga reptans
Labiate family (Lamiaceae)
MAY–JUNE 10–30CM

Note Spreads by leafy runners.
Description Inflorescence a leafy spike. Stem square, hairy along two sides. Leaves undivided.
Distribution Nutrient-rich, somewhat damp loamy soils to 2000m. Meadows, scrub and footpaths. Common, but rarer above 1500m.

Northern Dragonhead

Dracocephalum ruyschiana
Labiate family (Lamiaceae)
JULY–AUG. 5–15CM ▼

Note Flowers blue, to 3cm long in terminal clusters.
Description Leaves narrow, linear-lanceolate, with prominent midrib and rolled edges. Upper stem leaves stalkless.
Distribution Dry, stony soils to 2200m. Open woods and mountain meadows. Very rare.

Zygomorphic flowers

Dragonmouth

Horminum pyrenaicum
Labiate family (Lamiaceae)
JUNE–AUG. 10–40CM

Note Violet flowers in one-sided spike.
Description Leaves long-stalked, wrinkled, with rounded teeth.
Distribution Nutrient-poor, calcareous soils between 1600 and 2900m. Meadows and pastures. Scattered.

Large Self-heal

Prunella grandiflora
Labiate family (Lamiaceae)
JUNE–AUG. 5–25CM ▼

Note Violet flowers in tight terminal cluster.
Description Lower lip of flower regularly-toothed. Flowerheads 1–5cm above upper pair of stem leaves.
Distribution Nutrient-poor, dry soils to 2000m. Open woods and dry grassland. Scattered.

Alpine Skullcap

Scutellaria alpina
Labiate family (Lamiaceae)
JUNE–SEPT. 10–30CM

Note Flowers grow above each other in 4 rows.
Description Flower bracts overlapping and often tinged violet. Stem hairy.
Distribution Stony, calcareous soils between 1500 and 2500m. Open grassland and rocky debris. Rare.

Zygomorphic flowers

Alpine Toadflax

Linaria alpina
Figwort family (Scrophulariaceae)
JUNE–SEPT. 5–25CM ▼

Note Unmistakable with its blue-violet and orange flowers – an unusual combination.
Description Leaves in whorls of 3–4, somewhat fleshy and frosted bluish.
Distribution Damp, stony, calcareous soils to 3000m. Rocky debris and stony grassland. Common.

Matted Globularia

Globularia cordifolia
Globularia family (Globulariaceae)
MAY–AUG. 5–10CM

Note Leaves heart-shaped.
Description Mat-forming growth. Flowerheads c. 1.5cm diameter. Leaves to 4cm long.
Distribution Calcareous, rocky soils to 2800m. Stony grassland, ridges, rocks and rocky debris. Scattered.

Bald-stemmed Globularia

Globularia nudicaulis
Globularia family (Globulariaceae)
JUNE–AUG. 5–25CM

Note Stem leafless.
Description Flowerheads c. 2cm diameter. Leaves to 15cm long and 3cm broad, widest at the tip, in individual rosettes.
Distribution Commonest in eastern Alps on calcareous soils to 2600m. Stony grassland, open woods and scrub. Scattered.

Zygomorphic flowers

Alpine Bartsia

Bartsia alpina
Figwort family (Scrophulariaceae)
JUNE–AUG. 5–20CM

Note Upper flower-coloured leaf-like bracts increase attractiveness to insects.
Description Plant covered in glandular hairs. Leaves opposite.
Distribution Damp, nutrient- and humus-rich soils. Snow-patches, spring-fed flushes and fens to 2500m. Rare.

Alpine Eyebright

Euphrasia alpina
Figwort family (Scrophulariaceae)
MAY–SEPT. 3–20CM

Note Flowers lilac and white, with a yellow patch on the lower lip.
Description Branching form the base. Flowers with dark stripes. Leaves oval, unstalked, with rough teeth.
Distribution Lime-poor meadows and pastures, especially in central Alps between 1800 and 2700m. Scattered.

Common Butterwort

Pinguicula vulgaris
Butterwort family (Lentibulariaceae)
MARCH–JUNE 5–15CM

Note Catches insects on fleshy, sticky leaves.
Description Flowers bluish-violet with a white patch on the lower lip, and a 3–6mm spur. Leaves yellowish-green, rolled slightly upwards at margin.
Distribution Nutrient-poor soils. Spring-fed flushes, wet rocks and bogs.

Zygomorphic flowers

Wood Scabious

Knautia dipsacifolia
Teasel family (Dipsacaceae)
JUNE–SEPT. 30–100CM

Note Flowering stems with hemispherical flowerheads.
Description Flowers violet, sometimes red, corolla 4-lobed. Stem with dense covering of bristly hairs. Leaves all undivided.
Distribution Damp, often shady soils to 2200m. Tall-herb communities and open woodland. Common.

Long-leaved Scabious

Knautia longifolia
Teasel family (Dipsacaceae)
JULY–AUG. 30–80CM

Note Leaves long and narrow, with pale midrib.
Description Flowerheads 4–6cm diameter. Flowers pinkish-violet. Upper stem hairy, lower stem glabrous.
Distribution Mainly in central eastern Alps to 2300m. Grazed grassland and tall-herb communities. Scattered.

Shining Scabious

Scabiosa lucida
Teasel family (Dipsacaceae)
JUNE–SEPT. 15–40CM

Note Leaves shiny on upper surface.
Description Flowerheads 2–4cm diameter, the outer florets larger than inner ones. Black calyx bristles between florets. Basal leaves oval-lanceolate, upper stem leaves pinnately-lobed.
Distribution Stony, mainly calcareous soils to 2200m. Grassland, scree and moraines. Scattered.

Four petals

Lesser Meadow-rue

Thalictrum minus
Buttercup family (Ranunculaceae)
MAY–AUG. 15–150CM ▼

Note Conspicuous yellow stamens.
Description Flowers with small yellow perianth segments, growing in a loose panicle. Leaves 3–4-ternate, almost fern-like.
Distribution Dry, stony soils. Meadows and open woods. Scattered.

Yellow Alpine Poppy

Papaver alpinum ssp. *rhaeticum*
Poppy family (Papaveraceae)
JULY–AUG. 5–20CM ▼

Note Flowers bright golden-yellow.
Description Flowers to 5cm across. Basal leaves simply pinnate, with blunt loves. Plant has milky sap.
Distribution Mainly south-western and southern Alps, to 3000m, calciphile. Rocky debris and moraines. Scattered.

Cypress Spurge

Euphorbia cyparissias
Spurge family (Euphorbiaceae)
MAY–JULY 15–45CM ☠

Note Leaves narrow and blue-green, in an almost conifer-like arrangement on the stem.
Description Flowers in umbel-like inflorescence. The very small flowers are surrounded by 2 yellow bracts. White milky sap.
Distribution In Alps to about 2500m. Dry soils. Grassland and scree. Scattered.

Four petals

Alpine Alison

Alyssum alpestre
Crucifer family (Brassicaceae)
JUNE–JULY 5–25CM

Note Small rich-yellow flowers in dense, compact racemes.
Description Mat-forming. Leaves felty-hairy, grey-green. Podlet (silicula) rounded, oval and hairy.
Distribution Western Alps, east to Valais, to 3000m. Rocks and scree. Rare.

Buckler Mustard

Biscutella laevigata
Crucifer family (Brassicaceae)
MAY–JULY 10–30CM

Note Fruit 2-lobed and flattened – rather like a pair of spectacles.
Description Flowers pale yellow. Basal leaves in rosette, narrowly-lanceolate, 3–12cm long, short-stalked. Stem leaves unstalked.
Distribution Nutrient-poor, lime-rich soils between 800 and 2800m. Grassland, pastures and rocky debris. Scattered.

Drooping Bittercress

Cardamine enneaphyllos
Crucifer family (Brassicaceae)
MAY–JULY 20–30CM

Note 3-lobed leaves, mainly in threes towards the tip of otherwise leafless stems.
Description Inflorescence nodding. Flowers 1–2cm long, yellowish-white, in an umbel-like raceme. Basal leaves appear after flowering.
Distribution Deciduous woods in eastern Alps, to 1600m, on limestone. Scattered.

Four petals

Nasturtium-leaved Hairy Rocket

Erucastrum nasturtiifolium
Crucifer family (Brassicaceae)
MAY–SEPT. 30–80CM

Note Rich-yellow flowers to more than 2.5cm diameter.
Description Calyx hairy, the lobes almost horizontal. Stem hairy towards base. Leaves pinnately-lobed, with 4–8 lobes on each side.
Distribution Damp, calcareous soils to 1900m. River valleys, riverbanks and lake shores. Scattered.

Tansy-leaved Rocket

Hugueninia tanacetifolia
Crucifer family (Brassicaceae)
JUNE–AUG. 20–100CM

Note Leaves rather fern-like, or like those of Tansy.
Description Umbel-like panicles. Flowers to 8mm diameter. Stem leafy, with fluffy hairs. Pods almost rectangular.
Distribution Western Alps to Valais, to 2500m. Damp, nutrient-rich soils. Tall-herb communities and sites enriched by livestock dung. Scattered.

Yellow Whitlowgrass

Draba aizoides
Crucifer family (Brassicaceae)
APRIL–AUG. 5–15CM ▼

Note Basal leaves form a domed rosette.
Description Flowers to 1cm across, in groups of 5–10. Stem leafless. Leaves linear and rather thick, hairy. Podlet (silicula) hairy.
Distribution Stony, calcareous soils. Rock crevices, scree and ridges, between 1500 and 3000m. Common in calcareous Alps, otherwise rare.

Four petals

Hoppe's Whitlowgrass

Draba hoppeana
Crucifer family (Brassicaceae)
APRIL–AUG. 1–5CM ▼

Note Rosette leaves stiff, keeled and with ciliate margins.
Description Flowers 1–5 on leafless stems. Leaves linear-lanceolate. Rosette almost spherical.
Distribution Stony, gritty soils, mainly on calcareous shale, between 1500 and 3000m. Snow-patches and scree. Rare.

Swiss Treacle-mustard

Erysimum rhaeticum
Crucifer family (Brassicaceae)
MAY–AUG. 10–50CM

Note Flowers strongly-scented.
Description Flowers 12–20mm diameter. Leaves narrowly-lanceolate, with appressed hairs above and below.
Distribution Western Alps, east to Adige (Etsch), rarer in the north. Stony, lime-poor soils. Dry grassland, rocks and scree. Scattered.

Roseroot

Rhodiola rosea
Stonecrop family (Crassulaceae)
JULY–AUG. 10–40CM ▼

Note Flowers mostly 4-partite, in dense umbels, often red towards tip.
Description Dioecious (male and female flowers on separate plants). Flowers sometimes without petals. Leaves flat, fleshy and toothed. Medicinal plant.
Distribution Stony, mainly lime-poor soils to 3000m. Moraines, rock crevices and scree. Scattered.

Four petals

Tormentil

Potentilla erecta
Rose family (Rosaceae)
MAY–SEPT. 15–30CM

Note Flowers with 4 petals, slightly notched.
Description Stamens numerous. Stem leaves trifoliate, unstalked. Plant used medicinally and as a flavouring for spirits.
Distribution Damp, humus-rich soils to 2500m. Common.

Yellow Veronica

Paederota lutea
Plantain family (Plantaginaceae)
MAY–JULY 5–10CM

Note Leaves nettle-like, but soft, with more than 10 teeth on each side.
Description Plant rather drooping. Flowers 4-tipped to 2-lobed. Flower stalk densely-hairy.
Distribution Only in south-eastern and north-eastern calcareous Alps to 2500m. Rock crevices and scree. Rare.

Yellow Cephalaria

Cephalaria alpina
Teasel family (Dipsacaceae)
JULY–AUG. 60–100CM ▼

Note The pale yellow, hemispherical flowerheads are unmistakable.
Description Flowerheads 2.5–5cm diameter. Petals hairy on outside. Leaves pinnate.
Distribution Western Alps, east to Vorarlberg, to 1800m. Tall-herb communities and scrub on calcareous soils. Rare.

Five petals

Marsh-marigold

Caltha palustris
Buttercup family (Ranunculaceae)
MARCH–JUNE 15–50CM ▼ ☠

Note Flowers shiny golden-yellow.
Description Flower stalks curve upwards. Leaves heart- or kidney-shaped, to 15cm diameter.
Distribution Throughout the Alps from lowlands to 2400m. Wet or peaty soils. Streamsides and flushes. Common.

Hybrid Buttercup

Ranunculus hybridus
Buttercup family (Ranunculaceae)
JULY–AUG. 5–15CM ▼ ☠

Note Basal leaves toothed like a cockscomb.
Description Flowers small, to 1.5cm diameter. Basal leaves usually 1–2, fleshy-leathery, often frosted.
Distribution Northern and southern ranges of eastern Alps to 2500m, calcicole. Snow-patches, scree and rock crevices. Scattered.

Lesser Spearwort

Ranunculus flammula
Buttercup family (Ranunculaceae)
JULY–OCT. 15–50CM ☠

Note All leaves narrowly-lanceolate and entire.
Description Growth form varies, branching above. Flowers pale yellow, 0.5–2cm diameter. Lower leaves long-stalked.
Distribution Wet soils subject to occasional drying. Meadows, pastures and mires. Common below 1500m, above this rarer.

Five petals

Mountain Buttercup

Ranunculus montanus
Buttercup family (Ranunculaceae)
MAY–AUG. 5–30CM ▼ ☠

Note Flowers large, golden-yellow and shiny.
Description Flowers to 3.5cm diameter, in groups of 1–3. Basal leaves palmately-lobed, central lobe unstalked.
Distribution Damp pastures, fens and snow-patches. Common in calcareous Alps, otherwise rarer.

Woolly Buttercup

Ranunculus lanuginosus
Buttercup family (Ranunculaceae)
JUNE–AUG. 30–70CM ☠

Note Whole plant very hairy.
Description Flowers shiny and gold. Stem branching and many-flowered. Leaves palmately-lobed and irregularly-toothed.
Distribution Between 1000 and 2000m on damp, humus-rich soils. Woods and tall-herb communities. Common.

Meadow Buttercup

Ranunculus acris
Buttercup family (Ranunculaceae)
MAY–SEPT. 30–100CM ☠

Note Basal leaves and stem leaves differently shaped.
Description Flowers 1–2.5cm diameter, golden-yellow and shiny. Basal leaves 3–7-divided.
Distribution Throughout the Alps to over 2000m. Damp, nutrient-rich loamy soils. Meadows, pastures and footpaths. Common.

Five petals

Creeping Buttercup

Ranunculus repens
Buttercup family (Ranunculaceae)
MAY–AUG. 15–60CM

Note Spreads by creeping and rooting runners.
Description Flowers 2–3cm diameter, golden-yellow, shiny. Flower stems furrowed, basal leaves 3-divided, with stalked central lobe.
Distribution Damp, nutrient-rich soils. Wet grassland, footpaths and weedy sites, to 2400m. Common.

Imperforate St John's-wort

Hypericum maculatum
St John's-wort family (Clusiaceae)
JUNE–AUG. 20–60CM

Note Stem square.
Description Flowers to 2cm diameter, red when rubbed. Sepals and petals with black dots (glands) on underside. Leaves without translucent dots.
Distribution Meadows and pastures, poor grassland and scrub, to 2300m. Common.

Common Rock-rose

Helianthemum nummularium
Rock-rose family (Cistaceae)
MAY–SEPT. 10–30CM

Note Leaves with stipules.
Description Flowers 2–3cm diameter. Stem trailing or erect. Leaves linear-lanceolate.
Distribution Throughout Alps with many subspecies, to 2700m. Dry, nutrient-rich soils. Grassland and scree. Scattered.

Five petals

Alpine Rock-rose

Helianthemum alpestre
Rock-rose family (Cistaceae)
JULY–AUG. 5–15CM

Note Leaves without stipules.
Description Cushion growth form. Flowers 1–2cm diameter. Stem 1–5-flowered. Leaves linear-lanceolate.
Distribution Mainly in calcareous Alpine regions to 2900m. Stony grassland, often in exposed sites. Scattered to rare.

Yellow Rock-jasmine

Androsace vitaliana
Primrose family (Primulaceae)
MAY–JULY 2–5CM ▼

Note Flowers short-stalked, golden-yellow, growing proud of cushion.
Description Corolla tube 1cm long. Leaves narrowly-lanceolate, with white margins and hairy beneath, in rosettes.
Distribution Mainly in western Alps. Prefers lime-poor soils to 3000m. Snow-patches and scree. Rare.

Auricula

Primula auricula
Primrose family (Primulaceae)
APRIL–JUNE 5–25CM ▼

Note Rosette leaves glabrous, fleshy and often mealy-white.
Description Flowers bright yellow with a white ring at the centre, in umbel-like cluster. Leaves with pale, rather tough margin.
Distribution Mainly in calcareous Alps to 2500m. Stony grassland, rocks and mires. Rare.

Five petals

Oxlip

Primula elatior
Primrose family (Primulaceae)
MARCH–MAY 10–30CM ▼

Note Flowers pale yellow with inner orange ring.
Description Calyx lies close to corolla tube. Leaves wrinkled, hairy on both surfaces. Ancient medicinal plant.
Distribution Nutrient-rich damp, loamy soils between 300 and 2300m. Woods, meadows and scrub. Common.

Cowslip

Primula veris
Primrose family (Primulaceae)
MARCH–APRIL 10–25CM ▼

Note Flowers golden-yellow with five orange marks on the inside.
Description Calyx inflated, not pressed close to corolla tube. Leaves glabrous above. Ancient medicinal plant.
Distribution Nutrient-poor, dry, calcareous soils between 400 and 2200m. Woods, meadows and scrub. Scattered.

Alpine Stonecrop

Sedum alpestre
Stonecrop family (Crassulaceae)
JUNE–AUG. 2–8CM ▼ (☠)

Note Flowers, at 5–10mm, very small, compared with those of other stonecrops.
Description Petals blunt. Produces non-flowering shoots. Leaves without spur.
Distribution Damp, stony soils between 1500 and 3500m. Snow-patches, rocks, scree, moraines. Common.

Five petals

Biting Stonecrop

Sedum acre
Stonecrop family (Crassulaceae)
JUNE–AUG. 5–15CM

Note Petals golden-yellow, and pointed.
Description Produces non-flowering shoots. Inflorescence small and umbel-like. Leaves have sharp taste. Fruits whitish and star-shaped.
Distribution Dry, stony, mainly calcareous soils to over 2000m. Rocks, grassland and walls. Common.

Annual Stonecrop

Sedum annuum
Stonecrop family (Crassulaceae)
JUNE–AUG. 5–15CM

Note All shoots produce flowers. Plant annual.
Description Petals narrow and pointed. Stem upright and branched. Sepals fleshy.
Distribution Mainly central Alps. Dry, stony, lime-poor soils, to 2800m. Rocks, scree and walls. Scattered.

Mountain Stonecrop

Sedum montanum
Stonecrop family (Crassulaceae)
JUNE–JULY 10–30CM

Note Leaves to 2cm long, with an awn-like tip.
Description Inflorescence umbel-like. Leaves with spur on underside. Sharp taste.
Distribution Western Alps, inner valleys on dry, stony soils, between 400 and 1600m. Common.

Five petals

Yellow Mountain Saxifrage

Saxifraga aizoides
Saxifrage family (Saxifragaceae)
JUNE–OCT. 3–30CM ▼

Note Leaves cylindrical and fleshy, with a short, pointed tip.
Description Flowers orange-yellow, petals often spotted. Leaves shiny and rigid.
Distribution Scattered. Damp, mostly calcareous, often seeping soils, to 3000m. Rocks, spring-fed flushes and streamsides. Common.

Moss Saxifrage

Saxifraga bryoides
Saxifrage family (Saxifragaceae)
JULY–AUG. 3–15CM ▼

Note Petals whitish-yellow with central yellow-orange spot.
Description Petals much longer than sepals. Leaves of flowering plants form moss-like rosettes.
Distribution Lime-poor, damp, stony soils, 1800–4000m. Rocks, scree and snow-patches. Common in central Alps.

Musky Saxifrage

Saxifraga moschata
Saxifrage family (Saxifragaceae)
JULY–AUG. 1–10CM ▼

Note Flowers 1–5 on leafy stems.
Description Forms dense cushions. Flowers yellow-green, rarely white. Basal leaves often 3-lobed, some leaves unlobed.
Distribution Damp, calcareous soils between 1200 and 4000m. Rock crevices and scree. Common.

Five petals

Seguier's Saxifrage

Saxifraga seguieri
Saxifrage family (Saxifragaceae)
JUNE–AUG. 2–5CM ▼

Note Petals small and pale yellow – narrower than sepals.
Description Forms loose, flat mats. Plant has glandular hairs. Basal leaves oval, entire.
Distribution From Savoie to the Dolomites. Lime-poor, damp rocky and stony soils to 3000m. Scattered.

Leafless Saxifrage

Saxifraga aphylla
Saxifrage family (Saxifragaceae)
JULY–AUG. 1–5CM ▼

Note Petals and the narrower sepals both yellow-green, so flowers appear 10-lobed.
Description Stem leafless. Leaves mostly 3-lobed towards tip. Plant with glandular hairs.
Distribution Northern and southern calcareous Alps to 3200m. Rock crevices and scree. Scattered.

Eastern Saxifrage

Saxifraga sedoides
Saxifrage family (Saxifragaceae)
JUNE–OCT. 3–20CM ▼

Note Basal leaves flat, with spiny tip.
Description Petals pale- to lemon-yellow, never reddish. Leaves with glandular hairs.
Distribution Eastern Alps on calcareous soils between 1600 and 2800m, mainly on exposed, northern screes. Scattered.

Five petals

Golden Cinquefoil

Potentilla aurea
Rose family (Rosaceae)
JUNE–SEPT. 5–25CM

Note Petals golden-yellow, overlapping, and with orange spots at base.
Description Petals shallowly-notched. Leaves palmately-lobed, edged with silky hairs.
Distribution Nutrient- and lime-poor meadows and pastures between 1200 and 2800m. Common.

Alpine Cinquefoil

Potentilla crantzii
Rose family (Rosaceae)
MAY–SEPT. 5–20CM

Note Stem hairy.
Description Petals shallowly-notched. Leaves palmately-lobed, leaflets toothed on both sides. Upper surface of leaves with appressed hairs, lower surface with erect hairs.
Distribution Stony, calcareous, dry soils between 1000 and 3000m. Grassland and ridges. Scattered.

Glacier Cinquefoil

Potentilla frigida
Rose family (Rosaceae)
JULY–AUG. 2–10CM

Note Flowers to maximum 12mm across and mostly not fully open.
Description Stem 1–3-flowered and, like the three-lobed leaves, densely covered with shaggy hairs.
Distribution Only in central Alps. Dry, lime-free soils between 2500 and 3500m. Rocks and scree. Scattered.

Five petals

Large-flowered Cinquefoil

Potentilla grandiflora
Rose family (Rosaceae)
JUNE–SEPT. 10–40CM

Note Stems hairy, branching and many-flowered. Flowers to 3cm across.
Description Leaves 3-lobed, on long hairy stalks. Leaflets oval, evenly round-toothed.
Distribution Lime-poor, dry, stony, loamy soils to 2500m. Poor grassland and rocky slopes. Scattered.

Sibbaldia

Sibbaldia procumbens
Rose family (Rosaceae)
JUNE–SEPT. 2–10CM

Note Petals narrow, smaller than the broad sepals.
Description Grass-like growth. Stem softly-hairy. Leaves 3-lobed, grey-green. Each lobe with 3 teeth.
Distribution Damp, lime-poor soils between 2000 and 2800m. Snow-patches and grassland. Scattered.

Three-veined Hare's-ear

Bupleurum ranunculoides
Umbellifer family (Apiaceae)
JULY–AUG. 5–30CM ▼

Note The 5 bracteoles beneath the umbels are yellow and extend beyond the umbels.
Description Basal leaves elongate-lanceolate, almost grass-like, < 5mm across and up to 10cm long.
Distribution From western Alps to Allgäu and South Tyrol, to 2400m. Dry, nutrient-poor, but lime-rich soils. Scattered.

Five petals

Smooth Honeywort

Cerinthe glabra
Borage family (Boraginaceae)
MAY–JULY 30–50CM

Note Corolla yellow and tubular with a ring formed of red spots and blunt lobes, recurved at the tip.
Description Inflorescence drooping. Leaves bluish.
Distribution Nutrient-rich, calcareous, damp soils to 2300m. Scrub, tall-herb communities. Rare.

Yellow Bellflower

Campanula thyrsoides
Bellflower family (Campanulaceae)
JULY–SEPT. 10–50CM ▼

Note Inflorescence an impressive conical spike of pale yellow flowers.
Description Whole plant is roughly-hairy. Leaves narrowly-lanceolate and entire, with wavy margins.
Distribution Dry, stony, calcareous soils to 2500m. Grassland and scree. Rare.

Alpine Valerian

Valeriana celtica
Valerian family (Valerianaceae)
JULY–AUG. 5–15CM ▼

Note Flowers to 3mm across, tinted reddish at edges.
Description Leaves narrowly-lanceolate and entire. Typical valerian smell. Ancient medicinal plant, also used in cosmetics.
Distribution Two subspecies. Southwestern and eastern Alps on lime-poor humus soils to 3000m. Rare.

Many petals

Yellow Alpine Pasqueflower

Pulsatilla alpina ssp. *apiifolia*
Buttercup family (Ranunculaceae)
MAY–AUG. 10–40CM ▼ ☠

Note Easily identified by the sulphur-yellow flowers.
Description Flowers 3–6cm diameter. Stem leaves not joined at base.
Distribution Mainly in central Alpine ranges, between 1400 and 2600m on lime-poor soils (poor grassland). Scattered to common.

Globeflower

Trollius europaeus
Buttercup family (Ranunculaceae)
MAY–JUNE 20–60CM ▼ (☠)

Note Flowers globular, the petal-like sepals incurved.
Description Flowers to 3.5cm, with 8–15 petal-like sepals. Basal leaves rounded, palmately-lobed.
Distribution Throughout the Alps on damp, humus-rich soils. Meadows, tall-herb communities and streamsides. Scattered to common.

Hen-and-chickens Houseleek

Sempervivum globiferum
Stonecrop family (Crassulaceae)
JULY–SEPT. 10–25CM ▼

Note Inflorescence dense and umbel-like, with pale yellow, fringed flowers.
Description Basal leaves fleshy, in rosettes 1–7cm across.
Distribution Lime-poor rocky sites, from 1500 to 2400m. A number of similar subspecies. Scattered.

Many petals

Wulfen's Houseleek

Sempervivum wulfenii
Stonecrop family (Crassulaceae)
JULY–AUG. 10–30CM ▼

Note Rosette leaves fleshy and blue-green, tipped with a tough spine.
Description Flowers with red stamens, in dense, terminal raceme. Leaves densely fringed with soft hairs.
Distribution Only in eastern Alps, to 2700m. Lime-poor soils. Stony grassland, rock crevices and scree. Rare.

Great Yellow Gentian

Gentiana lutea
Gentian family (Gentianaceae)
JUNE–AUG. 40–140CM ▼

Note Golden-yellow flowers in dense, whorled clusters, each with 5–8 lobes, split almost to the base.
Description Leaves opposite. The similar, poisonous, White False Helleborine (p. 92) has alternate leaves. Medicinal plant, extract also used to flavour spirits.
Distribution Lime- and nutrient-rich mountain meadows to 2500m. Scattered.

Spotted Gentian

Gentiana punctata
Gentian family (Gentianaceae)
JUNE–SEPT. 20–70CM ▼

Note Flowers pale yellow with dark spots, bell-shaped, with 5–8 lobes.
Description Leaves opposite, shiny on upper surface.
Distribution Lime-poor, deep soils, 1500–3000m. Grassland and dwarf-shrub heath. Scattered, but absent from north-eastern calcareous Alps.

Many petals

Alpine Avens

Parageum montanum
Rose family (Rosaceae)
MAY–JULY 5–40CM

Note Rosette leaves pinnate, the terminal leaflet distinctly larger than the other leaflets.
Description Plant without runners. Petals 5–6, golden-yellow; sepals green.
Distribution Nutrient-poor meadows, pastures and dwarf-shrub heath, to 2800m. Common.

Creeping Avens

Geum reptans
Rose family (Rosaceae)
JULY–AUG. 5–15CM ▼

Note Flowers mostly solitary, with 5–7 golden-yellow petals.
Description Spreads by long overground runners. Rosette leaves pinnate, the leaflets deeply-toothed.
Distribution Damp lime-poor soils, 2000–2800m. Scree and moraines. Common in central Alps, otherwise rare.

Yellow Melancholy Thistle

Cirsium erisithales
Composite family (Asteraceae)
JUNE–AUG. 30–120CM

Note Inflorescence long-stalked, with sticky involucral bracts.
Description Flowers lemon-yellow, in nodding flowerheads. Leaves soft, with 8–12 spiny-toothed lobes.
Distribution Lime- and nutrient-rich mountain meadows to 2500m. Scattered.

Many petals

Spiniest Thistle

Cirsium spinosissimum
Composite family (Asteraceae)
JULY–SEPT. 20–80CM

Note Flowerheads pale yellow, surrounded by sharply-pointed, spiny leaves.
Description Plant very spiny. Flowerheads in dense cluster.
Distribution Damp, nutrient-rich soils to 2000m. Open woods and scrub. Scattered.

Berardia

Berardia subacaulis
Composite family (Asteraceae)
JUNE–AUG. 5–15CM

Note Flowerheads solitary, to 7cm across, in leafy rosette.
Description Stem very short or absent. Leaves oval, wrinkled and densely felty-hairy.
Distribution Only in south-western Alps. Stony, calcareous soils between 1500 and 2800m. Rocks, scree. Rare.

Arnica

Arnica montana
Composite family (Asteraceae)
MARCH–APRIL 30–60CM ▼

Note Stem with only 1–2 opposite pairs of leaves.
Description Flowerheads 6–8cm diameter, solitary. Basal leaves in rosette. Medicinal plant.
Distribution Nutrient- and lime-poor meadows and mires to 2800m. Scattered.

Many petals

Colts-foot

Tussilago farfara
Composite family (Asteraceae)
MARCH–JUNE 10–30CM

Note Outer ray florets narrow and numerous.
Description Flowerheads solitary on thick, scaly-leaved stems. Basal leaves appear after flowering.
Distribution Pioneer plant, to 2500m. Rocky scree, banks and waste land. Scattered.

Large-flowered Leopard's-bane

Doronicum grandiflorum
Composite family (Asteraceae)
MAY–AUG. 10–50CM ▼

Note Stem leaves clasping the stem.
Description Flowerheads 4–8cm across. Rows of flower bracts. Stem hairy. Basal leaves oval. Leaf margins hairy and coarsely-toothed.
Distribution Damp, calcareous to 1900m. Scree, rocks, and snow-patches. Scattered.

Tufted Leopard's-bane

Doronicum clusii
Composite family (Asteraceae)
JULY–AUG. 10–30CM

Note Basal leaves and stem leaves not heart-shaped or clasping the stem.
Description Leaves thin, almost glabrous on upper surface, with frizzy hairs at margin.
Distribution Lime-poor, stony soils with long snow-cover, between 1800 and 2500m. Scree, rock crevices and snow-patches. Scattered.

Many petals

Glacier Leopard's-bane

Doronicum glaciale
Composite family (Asteraceae)
JUNE–SEPT. 10–40CM

Note Flowerheads to 5cm across, terminal.
Description Flower bracts narrow, in a single row. Basal leaves elongate, stalked, with a few teeth. Upper stem leaves clasping the stem.
Distribution Stony grassland and snow-patches, between 1600 and 2600m. Rare.

Chamois Ragwort

Senecio doronicum
Composite family (Asteraceae)
JULY–AUG. 20–50CM ▼

Note Leaves coarsely-toothed and with grey-felty hairs on underside.
Description Flowerheads 4–6cm diameter, long-stalked. Leaves alternate (Arnica, similar, has opposite leaves).
Distribution Calcareous, stony soils to 3000m. Pastures, heath and rock crevices. Scattered.

Grey Alpine Groundsel

Jacobaea incana
Composite family (Asteraceae)
JULY–SEPT. 5–15CM

Note Leaves and stem densely covered with white-felty hairs.
Description Flowerheads 3–15, in dense terminal umbel-like cluster. Ray florets 3–5. Leaves pinnate, with deeply-cut lobes.
Distribution Lime-free, dry, stony soils to 3000m. Scree and rocky slopes. Scattered.

Many petals

Rock Ragwort

Senecio rupestris
Composite family (Asteraceae)
JULY–AUG. 20–60CM

Note Plant has unpleasant smell. NB This is now regarded as a subspecies of Oxford Ragwort (Senecio squalidus ssp. rupestris).
Description Outer flower bracts tipped with black hairs. Leaves cut beyond midway. Leaflets toothed, hairy beneath.
Distribution Stony, nutrient-rich soils between 600 and 2400m. Footpaths and in areas where livestock gather. Scattered.

Alpine Ragwort

Senecio alpinus
Composite family (Asteraceae)
JUNE–SEPT. 20–120CM

Note Leaves entire, heart-shaped and woolly-hairy beneath.
Description Flowerheads 6–20 in an umbel-like inflorescence. Stem angled.
Distribution Damp, nutrient-rich calcareous soils. Areas where livestock gather, and stream sides, 800 to 2000m. Scattered.

Southernwood Ragwort

Jacobaea abrotanifolia
Composite family (Asteraceae)
JULY–SEPT. 10–40CM ▼

Note Flowerheads orange-yellow–orange-red, 2.5–4cm across.
Description Flower bracts exactly 21. Lobes of stem leaves narrow, linear and shiny.
Distribution Dry, stony soils between 1400 and 2600m. Dwarf-shrub heath, grassland and Mountain Pine scrub. Rare.

Many petals

Gaudin's Ragwort

Tephroseris longifolia ssp. *gaudinii*
Composite family (Asteraceae)
JUNE–AUG. 30–80CM ▼

Note Leaves with cobweb-like hairs above and below.
Description Flowerheads 2.5–4cm diameter, with usually 13 green flower bracts. Fruit densely-hairy.
Distribution Only in eastern Alps, to 2500m. Calcareous, nutrient-rich soils. Sites enriched by livestock dung, pastures and scrub. Rare.

Wormwood

Artemisia absinthium
Composite family (Asteraceae)
JULY–SEPT. 40–120CM

Note Whole plant covered in grey-white felty hairs and has bitter taste.
Description Flowerheads rounded, short-stalked and nodding. Leaves 2–3-pinnate with silky hairs on both sides. Ancient medicinal plant; also used to flavour spirits.
Distribution Dry, stony tracks and grassland, to 2000m. Rare.

Glacier Wormwood

Artemisia glacialis
Composite family (Asteraceae)
JULY–AUG. 5–15CM ▼

Note Flowerheads clustered 3–10, terminally on upright stalks.
Description Whole plant has aromatic scent. Stem and leaves with grey-felty hairs. Leaves stalked and divided.
Distribution Lime-poor, stony soils between 2200 and 3000m. Rock crevices and scree. Rare.

Many petals

Spiked Wormwood

Artemisia genipi
Composite family (Asteraceae)
JULY–SEPT. 5–15CM ▼

Note Stems leafy throughout length, with 5–30 yellow flowerheads.
Description Inflorescence nodding at first. Plant aromatic, with grey-felty hairs.
Distribution Stony, calcareous soils between 2400 and 3000m. Rock crevices and scree. Rare.

Alpine Wormwood

Artemisia umbelliformis
Composite family (Asteraceae)
AUG.–SEPT. 10–30CM ▼

Note Inflorescence not one-sided, with 3–20 long-stalked flowerheads.
Description Leaves 2-ternate, with grey-felty hairs. Ancient medicinal plant; also used in liqueurs.
Distribution Stony, lime-poor soils between 1500 and 3200m. Moraines, rock crevices and scree. Rare.

Goldenrod

Solidago virgaurea ssp. *alpestris*
Composite family (Asteraceae)
JULY–SEPT. 10–30CM

Note Flowerheads 15–20mm across in terminal racemes.
Description Leaves narrowly-lanceolate, 4–6 times as long as broad. Ancient medicinal plant.
Distribution Stony, lime-poor soils between 1300 and 2700m. Grassland, pastures and dwarf-shrub heath. Common.

Many petals

Splendid Heartleaf Ox-eye

Telekia speciosissima
Composite family (Asteraceae)
JULY–AUG. 15–60CM ▼

Note Flowerheads very large, to 6cm across, the outer florets 1mm broad.
Description Stems single-headed. Upper leaves clasping stem.
Distribution Only between Lake Lugano and Lake Garda, to 2000m. Stony, dry, calcareous soils. Rare.

Yellow Ox-eye

Buphthalmum salicifolium
Composite family (Asteraceae)
JULY–AUG. 15–50CM

Note Flowerheads to 6cm across, the outer florets 2–3mm broad.
Description Stem often branching from low down. Upper leaves unstalked, narrowly-lanceolate, entire.
Distribution Stony, loamy, calcareous soils. Dry grassland and scrub, to 2200m. Scattered to rare.

Giant Cat's-ear

Hypochoeris uniflora
Composite family (Asteraceae)
JULY–AUG. 30–50CM

Note Flowerheads 5–7cm across, consisting only of ray florets.
Description Flowerheads solitary. Stem thickening towards the top, stiffly-hairy, with 2–7 leaves. Rosette leaves toothed.
Distribution Lime-poor, humus soils, between 1500 and 2500m. Grassland, dwarf-shrub heath and scrub. Scattered.

Many petals

Swiss Hawkbit

Scorzoneroides helvetica
Composite family (Asteraceae)
JULY–AUG. 10–30CM

Note Flowerheads 2–3cm broad, drooping before opening.
Description Leaves with rounded teeth, mostly glabrous. Stem with 2–4 scale-like leaves.
Distribution Slightly damp, lime-poor, humus soils to 2500m. Rough grassland, dwarf-shrub heath and scrub. Common in central Alps.

Rough Hawkbit

Leontodon hispidus
Composite family (Asteraceae)
JUNE–AUG. 8–50CM

Note Flowerheads droop before opening.
Description Flowerheads 2–3cm diameter. Flower bracts with white hairs. Upper stem thickening slightly. Maximum of 3 scale-like stem leaves. Leaves roughly-hairy.
Distribution Meadows and pastures between 300 and 2700m. Common.

Mountain Hawkbit

Leontodon montanus
Composite family (Asteraceae)
JULY–SEPT. 3–15CM

Note Outer flower bracts with black hairs.
Description Flowerheads solitary, 2–4cm diameter, on leafless stems. Leaves with rounded teeth.
Distribution Slightly damp, calcareous, stony soils between 1800 and 2700m. Scattered.

Many petals

Alpine Dandelion

Taraxacum sect. *Alpina*
Composite family (Asteraceae)
JUNE–SEPT. 5–20CM

Note Outer flower bracts at base of flower-head short, 5–7mm.
Description Leaves with rounded teeth or divided. Milky sap. Fruit stem short, 3–5mm.
Distribution Damp, nutrient-rich soils. Meadows and scree, between 1800 and 2400m. A complex, apomictic (form of asexual reproduction) group, with many species and forms. Common.

Large-headed Hawk's-beard

Crepis conyzifolia
Composite family (Asteraceae)
JULY–AUG. 20–70CM

Note Basal leaves still present at flowering time.
Description Flowerheads 3–5cm diameter, 1–9 on the stem. Flower bracts with glandular hairs. Basal leaves round-toothed and hairy.
Distribution Lime-poor, humus soils between 1400 and 2700m. Meadows and pastures. Scattered.

Pyrenean Hawk's-beard

Crepis pyrenaica
Composite family (Asteraceae)
JUNE–AUG. 25–50CM

Note Basal leaves withered by flowering time.
Description Flowerheads 3–5cm diameter. Stem leaves round-toothed and hairy.
Distribution Damp, nutrient- and lime-rich soils. Meadows and tall-herb communities, to 2200m. Scattered.

Many petals

Alpine Hawk's-beard

Crepis alpestris
Composite family (Asteraceae)
JUNE–AUG. 10–40CM

Note Flowerheads 3–4cm across, on hairy stems, mostly solitary.
Description Flower bracts with grey-felty hairs. Basal leaves deeply-toothed.
Distribution Eastern Alps. Dry, stony, calcareous soils to 2200m. Open woods, meadows and pastures. Scattered.

Kerner's Hawk's-beard

Crepis jacquinii ssp. *kerneri*
Composite family (Asteraceae)
JULY–AUG. 5–15CM

Note Flower bracts covered with long, black hairs.
Description Leaves entire at first, later toothed. Stem leaves divided.
Distribution Damp, nutrient- and lime-rich soils between 1000 and 2200m. Meadows and tall-herb communities. Scattered.

Triglav Hawk's-beard

Crepis terglouensis
Composite family (Asteraceae)
JULY–AUG. 5–7CM ▼

Note Flowerhead to 5cm across. Stem very short.
Description Flower bracts with black hairs. Leaves divided, with triangular segments.
Distribution Stony, calcareous soils between 2000 and 2700m. Rock crevices and scree. Rare.

Many petals

Pygmy Hawk's-beard

Crepis pygmaea
Composite family (Asteraceae)
JULY–AUG. 5–15CM

Note Leaves mostly tinted violet on the underside.
Description Flowerheads 1.5–2.5cm across. Lower stem leaves heart-shaped, and long-stalked. Upper leaves divided, with large terminal lobe.
Distribution South-western Alps. Calcareous scree between 2000 and 2700m. Rare.

Mouse-ear Hawkweed

Hieracium pilosella
Composite family (Asteraceae)
JUNE–SEPT. 5–30CM

Note Spreads by stolons to form dense carpets.
Description Flowerheads to 2cm across. Leaves entire, with long hairs above and white-felty hairs below.
Distribution Dry, nutrient-rich soils between 300 and 3000m. Meadows and pastures. Common.

Chicory-leaved Hawkweed

Hieracium intybaceum
Composite family (Asteraceae)
JULY–AUG. 5–30CM

Note Flowers yellowish-white.
Description Flowerheads 2–4cm across. Basal leaves absent. Leaves narrowly-lanceolate, with 1–6 long teeth; sticky with glandular hairs.
Distribution Stony, lime-poor, dry soils between 1600 and 2700m. Grassland and scree. Scattered.

Many petals

Alpine Hawkweed

Hieracium alpinum
Composite family (Asteraceae)
MAY–OCT. 5–25CM

Note Flower bracts with long hairs.
Description Flowerheads solitary. Basal leaves lanceolate and entire; hairy, especially around edges.
Distribution Lime-poor, humus soils between 1700 and 3000m. Open grassland and dwarf-shrub heath. Common to scattered.

Silver Hawkweed

Hieracium villosum
Composite family (Asteraceae)
JULY–AUG. 15–40CM

Note Plant covered in long, shaggy, silvery hairs.
Description Stem with 1–4 flowerheads, each 3–5cm diameter. Flower bracts spreading. Leaves grey-green with wavy edges.
Distribution Stony, calcareous, warm soils between 1000 and 2500m. Rock crevices and scree. Scattered.

German Asphodel

Tofieldia calyculata
Lily family (Liliaceae)
MAY–JULY 10–30CM ▼

Note Flowers in terminal, spike-like racemes, to 6cm long.
Description Flowers to 1cm diameter. Stem almost leafless. Leaves linear, grass-like and pointed.
Distribution Nutrient-poor, usually calcareous fens and wet meadows to over 2000m. Scattered.

Many petals

Liotard's Star-of-Bethlehem

Gagea liotardii
Lily family (Liliaceae)
MAY–JULY 5–15CM ▼

Note Flowers star-shaped, 1–5 at end of leafless stem.
Description Flower stalks hairy. Basal leaves tubular and hollow, with ragged hairs.
Distribution Lime-poor, damp, nutrient-rich soils. Mountain meadows between 1000 and 2400m. Scattered to rare.

Wild Tulip

Tulipa sylvestris ssp. *australis*
Lily family (Liliaceae)
MAY–JULY 10–40CM ▼

Note Flowers yellow, the three outer tepals tinged red. This is the southern Alpine subspecies.
Description Stem glabrous, with 2–3 grey-green, lanceolate, fleshy leaves.
Distribution Only in the south-western and southern Alps. Mountain meadows between 800 and 2000m. Scattered.

Wild Daffodil

Narcissus pseudonarcissus
Lily family (Liliaceae)
MARCH–MAY 10–30CM ▼ ☠

Note Typical daffodil – the ancestor of garden forms.
Description Flowers pale yellow, with darker trumpet-shaped, 2–4cm long corona. Leaves as long as flower stalk.
Distribution Lime-poor, humus soils, between 800 and 1500m. Mountain meadows. Scattered.

Zygomorphic flowers

Northern Wolfsbane

Aconitum lycoctonum
Buttercup family (Ranunculaceae)
JUNE–AUG. 40–200CM ▼ ☠

Note Flowers pale yellow; helmet taller than wide.
Description Leaves palmately 5–7-lobed. Yellow Wolfsbane (Aconitum anthora) is similar, but its leaves are more finely divided. Both species are very poisonous!
Distribution Damp humus soils. Tall-herb communities and riverbanks. Mostly common.

Yellow Corydalis

Corydalis lutea
Poppy family (Papaveraceae)
MAY–SEPT. 10–30CM ▼

Note The only yellow corydalis (in the region).
Description Flowers, to 2cm long, with a backward-pointing spur, in a many-flowered raceme. Leaves pinnate, with oval, lobed leaflets.
Distribution Originally only in the southern calcareous Alps, to 1800m, but sometimes naturalized.

Yellow Wood Violet

Viola biflora
Violet family (Violaceae)
MAY–AUG. 5–15CM

Note Leaves heart- or kidney-shaped.
Description Flowers yellow, to 1.5cm diameter, streaked brown. Leaves with rounded teeth, stipules entire.
Distribution To 2500m on nutrient-rich, calcareous soils. Woods, scrub and tall-herb communities. Scattered.

Zygomorphic flowers

Mountain Pansy

Viola lutea
Violet family (Violaceae)
JUNE–AUG. 5–10CM ▼

Note All flowers yellow, with brown streaks at the centre.
Description Spreads by rhizomes. Flowers to 4cm long. Leaves narrow and lanceolate, bluntly-toothed, with large stipules.
Distribution West and central Switzerland and in Styria, to 2000m. Dry, nutrient-poor soils. Rare.

Wild Pansy

Viola tricolor
Violet family (Violaceae)
MAY–OCT. 10–25CM

Note Flowers violet-blue, or violet-blue, yellow and white, sometimes entirely yellow.
Description Flowers to 3cm long, with a dark yellow spot and dark streaks in the centre. Leaves lanceolate and bluntly-toothed.
Distribution To 2100m in the Alps. Fields, meadows and tracks. Scattered to common.

Brown Clover

Trifolium badium
Legume family (Fabaceae)
JULY–AUG. 10–30CM

Note Flowerheads turn brown from the base upwards as they age.
Description Flowerheads globose to oval, to 1.5cm diameter. Upper leaves almost opposite.
Distribution Nutrient-rich, calcareous soils to 3000m. Pastures, scree and loose rocks. Scattered.

Zygomorphic flowers

Alpine Bird's-foot Trefoil

Lotus alpinus
Legume family (Fabaceae)
JUNE–AUG. 5–20CM

Note Flowers terminal, in groups of 1–3, often with reddish tinge.
Description Tip of keel red. Stem angular. Leaves 5-lobed, 3 lobes short-stalked.
Distribution Between 1800 and 2800m on stony soils. Pastures, grassland and scree. Common.

Common Bird's-foot Trefoil

Lotus corniculatus
Legume family (Fabaceae)
MAY–SEPT. 5–40CM

Note Flowers in clusters of 3–8.
Description Tip of keel pale yellow. Stem angular. Leaves 3-lobed with 2 stipule-like leaflets at base of leaf stalk.
Distribution Stony, loamy soils. Meadows, well-drained grassland, scrub and rocks, to 2300m. Common.

Small Scorpion Vetch

Coronilla vaginalis
Legume family (Fabaceae)
APRIL–JULY 10–25CM

Note Leaves pinnate, the 5–13 leaflets blue-green and fleshy.
Description Flowers about 1cm long, 3–10 in a loose inflorescence. Leaves with large leaflet-like stipules.
Distribution Dry, stony, calcareous soils to 2200m. Rocky slopes and pinewoods. Rare.

Zygomorphic flowers

Alpine Kidney Vetch

Anthyllis vulneraria ssp. *alpestris*
Legume family (Fabaceae)
MAY–SEPT. 5–30CM

Note Flowers in dense heads, surrounded by palmate, leaf-like bracts.
Description Sepals pale yellow, petals golden-yellow. Leaves pinnate, with large terminal leaflet.
Distribution Calcareous, stony pastures and grassland to 2800m. Common.

Mountain Lentil

Astragalus penduliflorus
Legume family (Fabaceae)
JULY–AUG. 15–40CM

Note Flowers drooping, in dense raceme arising from a leaf axil.
Description Leaves pinnate, with 11–27 leaflets. Fruit pendent, inflated.
Distribution Stony, nutrient-poor soils between 1300 and 2600m. Meadows and open woodland. Rare.

Yellow Oxytropis

Oxytropis campestris
Legume family (Fabaceae)
JUNE–AUG. 5–20CM

Note Flowers 8–20, upright in rounded flowerheads.
Description Flowers to 2cm long. Leaves in basal rosette, pinnate with 21–31 leaflets.
Distribution Nutrient-poor, usually calcareous soils between 1600 and 3000m. Stony grassland and ridges.

Zygomorphic flowers

Horseshoe Vetch

Hippocrepis comosa
Legume family (Fabaceae)
MAY–JULY 5–30CM

Note Flowers in umbel-like terminal clusters of 3–10.
Description Leaves pinnate with 11–17 obovate leaflets.
Distribution Calcareous soils to 2200m. Dry, stony grassland. Scattered to rare.

Yellow Pea

Lathyrus laevigatus
Legume family (Fabaceae)
JUNE–AUG. 20–60CM

Note Plant without tendrils. The pinnate leaves end in a point, not a tendril.
Description Flowers fade to brown with age. Inflorescence 3–12-flowered. Leaves with 8–12 leaflets.
Distribution Calcareous soils to 2000m. Meadows and tall-herb communities. Very rare.

Meadow Vetchling

Lathyrus pratensis
Legume family (Fabaceae)
JUNE–JULY 20–80CM

Note Clambers using tendrils at tips of paired leaflets.
Description Flowers 3–13, in long-stalked racemes. Stipules leaf-like. Pods usually glabrous.
Distribution Damp, humus soils between 300 and 2000m. Meadows, edges of woods and tracks. Common.

Zygomorphic flowers

Shrubby Milkwort

Polygala chamaebuxus
Milkwort family (Polygalaceae)
APRIL–JUNE 10–25CM

Note Flowers two-toned – white and yellow (or pink and yellow).
Description Flowers to 1.5cm long. Stem woody at base. Leaves lanceolate, leathery.
Distribution Warm, dry, calcareous soils to 2500m. Poor grassland and dry woods. Scattered in northern Alps, otherwise very rare.

Yellow Archangel

Lamium galeobdolon
Labiate family (Lamiaceae)
MAY–JUNE 15–50CM

Note Easily identified by the nettle-like leaves and yellow flowers.
Description Flowers in whorls up the stem, each with 8–16 flowers. Flowers 20–25mm long. Leaves opposite.
Distribution Nutrient-rich, calcareous, damp soils to 2000m. Deciduous woods and scree. Scattered.

Sticky Sage

Salvia glutinosa
Labiate family (Lamiaceae)
JULY–SEPT. 40–80CM

Note Leaves large, pointed and sticky.
Description Flowers, 3–4cm long, in up to 10 whorls. Leaves opposite.
Distribution Nutrient-rich, calcareous soils to 1700m. Woods, scrub and tall-herb communities. Scattered in calcareous Alps, otherwise very rare.

Zygomorphic flowers

Hyssop-leaved Mountain Tea

Sideritis hyssopifolia
Labiate family (Lamiaceae)
JULY–SEPT. 10–40CM

Note Flowers pale yellow, in 5–15, 4–6-flowered whorls.
Description Flower bracts with spiny awns. Stem hairy. Leaves narrowly-lanceolate.
Distribution Calcareous, rocky and grassy slopes in the south-western Alps as far as the Swiss Jura. Scattered.

Yellow Betony

Stachys alopecuros
Labiate family (Lamiaceae)
JUNE–SEPT. 20–50CM

Note Flowers in a rather short, broad terminal spike composed of several whorls.
Description Upper lip of flower two-lobed. Leaves long-stalked, heart-shaped, with rounded teeth.
Distribution Stony soils between 500 and 1800m. Open woods, rocky slopes and meadows. Scattered.

Large Yellow Foxglove

Digitalis grandiflora
Figwort family (Scrophulariaceae)
JUNE–AUG. 50–130CM ▼ ☠

Note Flowers pale yellow, 3–4.5cm long, broadening to 1.5–2cm.
Description Flowers with brownish veins on the inside. Basal leaves form a rosette. Leaves and fruits hairy.
Distribution Stony, calcareous, nutrient-rich soils to 1900m. Woodland clearings, pastures and scrub. Scattered.

Zygomorphic flowers

Small Yellow Foxglove

Digitalis lutea
Figwort family (Scrophulariaceae)
JUNE–JULY 50–100CM ▼ ☠

Note Flowers pale yellow, 2–2.5cm long, broadening to 5–8mm.
Description Flowers with shaggy hairs, inside and outside. No basal leaf rosette.
Distribution Stony, calcareous soils to 1800m. Mountain woods and dwarf-shrub heath. Rare.

Dwarf Eyebright

Euphrasia minima
Figwort family (Scrophulariaceae)
JUNE–SEPT. 2–12CM ▼

Note Flowers yellow or white, often with violet upper lip.
Description Stem usually unbranched, hairy. Leaves oval, unstalked, with 6–14 teeth.
Distribution Nutrient-poor, usually also lime-poor meadows and pastures, between 1500 and 3200m. Common.

Common Cow-wheat

Melampyrum pratense
Figwort family (Scrophulariaceae)
JUNE–AUG. 15–50CM

Note Flowers in one-sided, leafy spikes.
Description Flowers pale yellow, 12–18mm long. Sepals short and pointed. Leaves lanceolate.
Distribution Nutrient- and lime-poor soils to 2000m. Woods and heath, mires, and sides of tracks. Scattered.

Zygomorphic flowers

Alpine Tozzia

Tozzia alpina
Broomrape family (Orobanchaceae)
JUNE–AUG. 15–50CM

Note Flowers golden-yellow, with reddish spots on the three lower lobes.
Description Flowers with 5 unequal lobes. Inflorescence leafy. Leaves with 1–3 teeth on each side.
Distribution Nutrient- and lime-rich soils to 2300m. Woods, scrub and tall-herb communities. Scattered to rare.

Awned Yellow-rattle

Rhinanthus glacialis
Figwort family (Scrophulariaceae)
MAY–SEPT. 15–50CM ▼ (☠)

Note Leaf-like involucral bracts sharply-toothed and with 1–3mm long awns.
Description Flowers to 2cm long. Calyx glabrous. Upper lip of flower with a blue tooth.
Distribution Periodically-damp, usually calcareous, but nutrient-poor soils between 800 and 2600m. Stony grassland. Scattered.

Greater Yellow-rattle

Rhinanthus alectorolophus
Figwort family (Scrophulariaceae)
MAY–SEPT. 10–50CM ▼ (☠)

Note Calyx slightly inflated and hairy.
Description Flowers to 2.5cm long. Upper lip with a blue tooth. Leaf-like involucral bracts sharply-toothed.
Distribution Periodically-damp, usually calcareous humus soils between 400 and 1900m. Meadows and spring-fed mires. Scattered.

Zygomorphic flowers

Leafy Lousewort

Pedicularis foliosa
Figwort family (Scrophulariaceae)
JUNE–AUG. 20–60CM ▼ (☠)

Note Inflorescence with long, leaf-like bracts.
Description Flowers pale yellow, in dense many-flowered racemes. Basal leaves 10–25cm long, 2-pinnate. All louseworts are semiparasitic.
Distribution Calcareous, stony soils to 2000m. Meadows and tall-herb habitats. Rare.

Crimson-tipped Lousewort

Pedicularis oederi
Figwort family (Scrophulariaceae)
JUNE–JULY 5–15CM ▼ (☠)

Note Flowers yellow, with crimson tip to upper lip.
Description Upper lip not beak-like. Calyx with long hairs. Leaves bluish-green, pinnately-lobed. Leaves toothed and overlapping.
Distribution Calcareous, damp soils to 2500m. Alpine grassland and heath. Rare.

Crested Lousewort

Pedicularis comosa
Figwort family (Scrophulariaceae)
JUNE–AUG. 10–50CM ▼ (☠)

Note Flowers pale- to lemon-yellow, in dense, broad, many-flowered spike.
Description Flowers large, 18–30mm long, without beak. Basal leaves long-stalked; upper stem leaves unstalked.
Distribution Western and southern Alps. Meadows and pastures to 2000m. Rare.

Zygomorphic flowers

Tuberous Lousewort

Pedicularis tuberosa
Figwort family (Scrophulariaceae)
JUNE–AUG. 10–20CM ▼ (☠)

Note Upper lip of flower is beak-like.
Description Lower stem and leaf stalks with long hairs. Leaves pinnately-lobed, the lobes deeply-toothed.
Distribution Lime- and nutrient-poor meadows and fens between 1500 and 2500m. Scattered to rare.

Pale Yellow Broomrape

Orobanche flava
Broomrape family (Orobanchaceae)
JUNE–JULY 15–40CM ▼

Note No green leaves – plant fully parasitic (mainly on Alpine Butterbur).
Description Flowers 1.5–2cm long, yellow or red. Stem pale yellow.
Distribution Calcareous scree habitats to 2000m. Common in northern calcareous Alps, otherwise scattered.

Lady's-slipper

Cypripedium calceolus
Orchid family (Orchidaceae)
MAY–JULY 20–70CM ▼

Note Unmistakable; flower has a large, pouched, slipper-like lip.
Description Flowers to 10cm diameter. Leaves 3–4 per stem, to 18cm long, with prominent veins and hairy beneath.
Distribution Northern and southern calcareous Alps to 2000m. Open woods and scrub. Scattered.

Zygomorphic flowers

Elder-flowered Orchid

Dactylorhiza sambucina
Orchid family (Orchidaceae)
APRIL–JUNE 15–30CM ▼

Note Stem leafy up to inflorescence.
Description Flowers pale yellow or purple, or bi-coloured. Flowers with downward-pointing spur, and long bracts. Lip spotted purplish.
Distribution Dry, lime-poor but nutrient-rich soils to 2100m. Meadows and open woods. Rare.

Pale-flowered Orchid

Orchis pallens
Orchid family (Orchidaceae)
APRIL–JUNE 15–30CM ▼

Note Leaves not reaching inflorescence.
Description Flowers always pale yellow. Spur horizontal or upward-pointing. Bracts membranous.
Distribution Moist humus-rich soils. Woods and mountain meadows. Mainly in calcareous Alps to 1700m. Scattered to rare.

Small White Orchid

Pseudorchis albida
Orchid family (Orchidaceae)
MAY–SEPT. 10–30CM ▼

Note Flowers small, to 0.5mm long, bell-shaped, in a long spike.
Description Lip 3-lobed, usually yellow. Spur short. Leaves 4–5, elongate.
Distribution Only in central Alpine ranges. Lime-free, humus soils between 1500 and 2500m. Scattered.

Four petals

Good King Henry

Chenopodium bonus-henricus
Goosefoot family (Chenopodiaceae)
JUNE–AUG 20–100CM ▼

Note Leaves triangular and pointed, to 15cm long.
Description Flowers inconspicuous. Inflorescence spike-like, often drooping. Young plant edible (like spinach).
Distribution Throughout Alps to 2500m. Nutrient-rich soils. Around Alpine huts, cattle sheds and footpaths. Scattered.

French Sorrel

Rumex scutatus
Dock family (Polygonaceae)
JUNE–JULY 20–50CM

Note Leaves spear-shaped, with spreading basal lobes.
Description Individual flowers greenish, in few-flowered, loose inflorescence. Leaves blue-green, long-stalked, to 5cm long.
Distribution All altitudes to 2200m. Dry, stony soils. Scree, rocky debris and rock crevices. Scattered.

Alpine Lady's-mantle

Alchemilla alpina
Rose family (Rosaceae)
JUNE–JULY 5–30CM

Note Leaves 5–9-lobed, divided almost to the base.
Description Leaflets narrowly-oval, toothed towards tip, and with silver silky sheen beneath. Several similar species.
Distribution Mostly acid, nutrient-poor, loamy soils to 2600m. Dwarf-shrub heath, Alpine grassland and rocky sites.

Four petals

Common Lady's-mantle

Alchemilla vulgaris agg.
Rose family (Rosaceae)
JUNE–JULY 5–40CM

Note Leaves 7–11-lobed, divided not quite to halfway.
Description Lobes rounded to triangular, toothed and green on both sides. This is an aggregate species name, within which several species have since been named.
Distribution Damp, nutrient-rich soils between 500 and 2800m. Meadows and pastures. Common.

Alpine Plantain

Plantago alpina
Plantain family (Plantaginaceae)
MAY–JULY 5–15CM

Note Leaves very narrow, almost grass-like, with 3 rather indistinct veins.
Description Flower spike 1.5–3cm long, cylindrical. Stalk of spike with appressed hairs. Flowers with yellow stamens.
Distribution Lime- and nutrient-poor soils between 1300 and 2500m. Scattered.

Dark Plantain

Plantago atrata
Plantain family (Plantaginaceae)
MAY–AUG. 5–15CM

Note Inflorescence spike rounded or oval.
Description Stamens whitish. Stalk of spike with short, erect hairs, not furrowed. Leaves narrow, distinctly 3–7-veined.
Distribution Nutrient-rich, calcareous, damp soils to 2400m. Meadows and pastures. Scattered.

Four/Five petals

Ribwort Plantain

Plantago lanceolata
Plantain family (Plantaginaceae)
MAY–OCT. 5–60CM

Note Flowers in oval to short-cylindrical spikes.
Description Stamens prominent, anthers yellowish. Stem with 5 longitudinal ridges. Rosette leaves narrow, 3–7-veined.
Distribution Nutrient-rich soils to 2000m. Meadows and pastures, also alongside tracks. Very common.

Greater Plantain

Plantago major
Plantain family (Plantaginaceae)
JUNE–OCT. 15–30CM

Note Flowers inconspicuous, in long spikes.
Description Rosette leaves longer than stalks of spikes. Anthers violet. The similar Hoary Plantain (Plantago media) has pale pink anthers.
Distribution Nutrient-rich, rather damp, loamy soils to 2500m. Tracks, pastures, and sites enriched by livestock dung. Very common.

Cyphel

Minuartia sedoides
Carnation family (Caryophyllaceae)
JULY–AUG. 2–5CM ▼

Note Forms dense, flat cushions.
Description Flowers yellow-green, small, to 0.5cm diameter, mostly without petals. Leaves 3–6mm long.
Distribution Mainly in the central ranges on lime-poor substrates to over 3500m. Stony grassland, rock crevices and scree. Common.

Five/Many petals/Zygomorphic flowers

Bilberry

Vaccinium myrtillus
Heath family (Ericaceae)
MAY–JUNE 15–50CM

Note Flowers bell-shaped to globular, often tinged red.
Description Stem green, square, and winged. Leaves elongate-oval. Fruits blue-black edible berries.
Distribution Acid, lime- and nutrient-poor soils to 2500m. Coniferous woods, dwarf-shrub heath and mires. Common.

Herb Paris

Paris quadrifolia
Lily family (Liliaceae)
MAY–JUNE 10–30CM

Note Single flower above four large leaves.
Description Perianth segments 6–12. Leaves in a whorl of 4, with network of veins. Single black berry-like capsule.
Distribution Nutrient- and humus-rich, damp soils to 1800m. Deciduous woods and streamsides. Scattered, commoner in valleys.

Broad-leaved Helleborine

Epipactis helleborine
Orchid family (Orchidaceae)
JUNE–SEPT. 20–70CM ▼

Note Flowers greenish, often violet-tinted, often on same side of inflorescence.
Description Lower lip of flower whitish to pale red. Leaves broadly-oval, at base clasping the stem.
Distribution Lime- and nutrient-rich soils in woods up to the tree-line at around 1700m. Scattered.

Zygomorphic flowers

Musk Orchid

Herminium monorchis
Orchid family (Orchidaceae)
JUNE–AUG. 10–30CM ▼

Note Flowers greenish-yellow, honey-scented, in a loose, cylindrical spike.
Description Flowers bell-shaped, with 3-lobed lip, angled downwards, without spur. Leaves mostly paired at base of stem.
Distribution Lime- and nutrient-poor damp meadows, to 1800m. Rare.

Lesser Twayblade

Listera cordata
Orchid family (Orchidaceae)
MAY–AUG. 5–20CM ▼

Note Flowers greenish, often with reddish tint. Lip of flower deeply-divided.
Description Flower without spur. Basal leaves 2, heart-shaped, 1–2.5cm diameter, almost opposite.
Distribution Acid and lime-poor soils with rich raw humus, to 2000m. Moss-rich coniferous woods. Rare.

Common Twayblade

Listera ovata
Orchid family (Orchidaceae)
MAY–JUNE 20–70CM ▼

Note Flowers with conspicuous long, two-lobed lip, growing in a long, spike-like raceme.
Description Flower without spur. Leaves oval, 5–15cm long, opposite.
Distribution Damp, mainly calcareous soils, to 2000m. Woods, meadows and mires.

Zygomorphic flowers

False Musk Orchid

Chamorchis alpina
Orchid family (Orchidaceae)
JULY–AUG. 5–15CM ▼

Note Plant small, with 5–10 flowers and grass-like leaves.
Description Flowers yellow-green, tinged red-brown on outside. Lip yellowish, usually entire or slightly 3-lobed.
Distribution Lime-rich, stony, dry soils above the tree-line, between 1700 and 3000m. Scattered.

Fly Orchid

Ophrys insectifera
Orchid family (Orchidaceae)
MAY–JULY 5–30CM ▼

Note Flowers resemble small, winged insects.
Description Flower with 3 greenish sepals. Lip brown and hairy, 4-lobed, with a grey-blue patch. Basal leaves upright.
Distribution Dry, calcareous soils, to 1800m. Open woods and poor grassland. Rare.

Frog Orchid

Dactylorhiza viridis
Orchid family (Orchidaceae)
MAY–JULY 5–30CM ▼

Note Outer tepals form a domed hood at top of flower.
Description Flowers often tinged brownish-red. Lip 3-lobed, with shorter middle lobe. Spur rounded.
Distribution Lime-poor, damp soils, to 2600m. Alpine grassland and snow-patches. Rare.

Index

Achillea atrata 88
– clavennae 88
– macrophylla 88
– millefolium 87
– moschata 89
– nana 89
Acinos alpinus 44
Aconitum lycoctonum 164
– napellus 123
– variegatum 123
Adenostyles 33
Adenostyles, Alpine 34
Adenostyles, Woolly 34
Adenostyles alliariae 33
– alpina 34
– leucophylla 34
Ajuga pyramidalis 127
– reptans 127
Alchemilla alpina 176
– vulgaris 177
Alison, Alpine 133
Allium narcissiflorum 40
– schoenoprasum 41
– victorialis 94
Alpenrose 19
Alpenrose, Dwarf 20
Alpenrose, Hairy 19
Alpine Bells 21
Alyssum alpestre 133
Androsace adfinis ssp.
  puberula 20
– alpina 20
– chamaejasme 68
– hausmannii 69
– helvetica 69
– lactea 68
– obtusifolia 69
– villosa 68
– vitaliana 140
Anemone baldensis 84
– narcissiflora 84
Anemone, Narcissus-flowered 84
Anemone, Tyrol 84
Angelica sylvestris 79
Angelica, Wild 79
Antennaria carpatica 87
– dioica 35
Anthericum liliago 93
Anthyllis vulneraria ssp.
  alpestris 167

Aquilegia alpina 106
– vulgaris 106
Arabis alpina 52
– caerulea 100
– pumila 52
– soyeri 53
Archangel, Yellow 169
Arctostaphylos alpinus 65
– uva-ursi 66
Arenaria biflora 64
– ciliata 60
Armeria alpina 18
Arnica 151
Arnica montana 151
Artemisia absinthium 155
– genipi 156
– glacialis 155
– umbelliformis 156
Asphodel, German 162
Asphodel, Scottish 91
Asphodel, White 91
Asphodelus albus 91
Aster, Alpine 120
Aster, False 89
Aster alpinus 120
– bellidiastrum 89
Astragalus alpinus 96
– australis 96
– frigidus 97
– penduliflorus 167
– sempervirens 97
Astrantia major 76
– minor 77
Athamanta cretensis 79
Auricula 21
Avens, Alpine 150
Avens, Creeping 150
Avens, Mountain 86
Azalea, Trailing 19

Bartsia, Alpine 130
Bartsia alpina 130
Basil-thyme, Alpine 44
Bastard-toadflax, Alpine 56
Bastard-toadflax, Pyrenean 76
Bearberry 66
Bearberry, Mountain 65
Bedstraw, Alpine 57
Bedstraw, Swiss 57
Bellflower, Alpine 117
Bellflower, Bearded 116

Bellflower, Clustered 117
Bellflower, Grass-like 119
Bellflower, Mount Cenis 118
Bellflower, Nettle-leaved 117
Bellflower, Scheuchzer's 119
Bellflower, Spiked 116
Bellflower, Yellow 147
Bellflower, Zois' 118
Berardia 151
Berardia subacaulis 151
Betony, Alpine 45
Betony, Yellow 170
Bilberry 179
Bilberry, Bog 67
Bird's-foot Trefoil, Alpine 166
Bird's-foot Trefoil, Common 166
Biscutella laevigata 133
Bistort, Alpine 65
Bistort, Common 18
Bistorta officinalis 18
– vivipara 65
Bitter-cress, Alpine 53
Bittercress, Drooping 133
Bitter-cress, Large 54
Bitter-cress, Mignonette-leaved 54
Bogbean 81
Braya alpina 53
Braya, Alpine 53
Broomrape, Pale Yellow 174
Bugle 127
Bugle, Pyramidal 127
Bugloss, Viper's 112
Buphthalmum salicifolium 157
Bupleurum ranunculoides 146
Burnet, Great 12
Burnet-saxifrage, Great 27
Butterbur, Alpine 34
Buttercup, Aconite-leaved 58
Buttercup, Alpine 59
Buttercup, Creeping 139
Buttercup, Glacier 58
Buttercup, Hybrid 137
Buttercup, Kuepfer's 59
Buttercup, Large White 58
Buttercup, Meadow 138
Buttercup, Mountain 138
Buttercup, Parnassus-leaved 59

Buttercup, Woolly 138
Butterfly-orchid, Lesser 99
Butterwort, Alpine 98
Butterwort, Common 130

Callianthemum, Coriander-leaved 84
Callianthemum coriandrifolium 84
Calluna vulgaris 31
Caltha palustris 137
Campanula alpina 117
– *barbata* 116
– *cenisia* 118
– *cespitosa* 119
– *cochleariifolia* 118
– *glomerata* 117
– *rhomboidalis* 119
– *scheuchzeri* 119
– *spicata* 116
– *thyrsoides* 147
– *trachelium* 117
– *zoysii* 118
Campion, Bladder 64
Campion, Moss 17
Campion, Red 17
Campion, Rock 63
Campion, Small 63
Candytuft, Rock 55
Caraway 80
Cardamine amara 54
– *bellidifolia* ssp. *alpina* 53
– *enneaphyllos* 133
– *pentaphyllos* 100
– *resedifolia* 54
Carduus defloratus 35
– *personata* 36
Carlina acaulis 86
Carrot, Candy 79
Carrot, Moon 80
Carum carvi 80
Cat's-ear, Giant 157
Cat's-foot, Carpathian 87
Catchfly, Alpine 18
Catchfly, Nottingham 64
Centaurea montana 121
– *nervosa* 37
– *phrygia* 37
– *scabiosa* ssp. *alpestris* 36
– *triumfettii* 121
– *uniflora* 36
Cephalaria alpina 136
Cephalaria, Yellow 136
Cerastium alpinum 60
– *arvense* 60
– *cerastoides* 61

– *latifolium* 61
– *uniflorum* 61
Cerinthe glabra 147
Chaerophyllum villarsii 81
Chamorchis alpina 181
Chenopodium bonus-henricus 176
Chervil, Alpine 81
Chives 41
Christmas Rose 85
Cicerbita alpina 122
Cinquefoil, Alpine 145
Cinquefoil, Clusius' 75
Cinquefoil, Dolomite 25
Cinquefoil, Glacier 145
Cinquefoil, Golden 145
Cinquefoil, Large-flowered 146
Cinquefoil, Shrubby White 75
Cirsium acaule 38
– *eriophorum* 38
– *erisithales* 150
– *helenioides* 38
– *spinosissimum* 151
Clematis alpina 100
Clematis, Alpine 100
Clover, Alpine 43
Clover, Brown 165
Clover, Mountain 95
Clover, Noble 43
Clover, Pale 95
Clover, Snow 95
Clover, Thal's 96
Colchicum alpinum 42
Colt's-foot, Purple 35
Colts-foot 152
Columbine, Alpine 106
Columbine, Common 106
Convallaria majalis 92
Coral-wort, Five-leaved 100
Cornflower, Perennial 121
Coronilla vaginalis 166
Cortusa matthioli 21
Corydalis lutea 164
Corydalis, Yellow 164
Cowberry 67
Cowslip 141
Cow-wheat, Common 171
Crane's-bill Alpine Wood, 76
Crane's-bill, Bloody 26
Crane's-bill, Dusky 26
Crane's-bill, Wood 26
Crepis alpestris 160
– *aurea* 40
– *conyzifolia* 159
– *jacquinii* ssp. *kerneri* 160
– *pygmaea* 161

– *pyrenaica* 159
– *terglouensis* 160
Cress, Chamois 56
Crowfoot, Glacier 58
Cyclamen, Alpine 21
Cyclamen purpurascens 21
Cyphel 178
Cypripedium calceolus 174

Dactylorhiza incarnata ssp. *cruenta* 49
– *maculata* 48
– *majalis* 48
– *sambucina* 175
– *viridis* 181
Daffodil, Wild 163
Dandelion, Alpine 159
Daphne alpina 57
– *cneorum* 12
– *striata* 12
Daphne, Alpine 57
Daphne, Striped 12
Dead-nettle, White 97
Delphinium elatum 106
Devil's Claw 28
Dianthus alpinus 15
– *barbatus* 15
– *glacialis* 15
– *superbus* 16
– *sylvestris* 16
Digitalis grandiflora 170
– *lutea* 171
Dock, Mountain 29
Dock, Snow 29
Doronicum clusii 152
– *glaciale* 153
– *grandiflorum* 152
Draba aizoides 134
– *hoppeana* 135
– *siliquosa* 54
– *tomentosa* 55
Dracocephalum ruyschiana 127
Dragonhead, Northern 127
Dragonmouth 128
Dryas octopetala 86

Echium vulgare 112
Edelweiss 87
Epilobium alpestre 14
– *alsinifolium* 13
– *angustifolium* 13
– *fleischeri* 13
Epipactis atrorubens 49
– *helleborine* 179
– *palustris* 99

Index

Erica carnea 11
Erigeron alpinus 39
– *uniflorus* 90
Erinus alpinus 113
Eritrichum nanum 112
Erucastrum nasturtiifolium 134
Eryngium alpinum 107
Eryngo, Alpine 107
Erysimum rhaeticum 135
Erythronium dens-canis 41
Euphorbia cyparissias 132
Euphrasia alpina 130
– *minima* 171
– *rostkoviana* 98
Eyebright, Alpine 130
Eyebright, Common 98
Eyebright, Dwarf 171

Fairy's Thimble 118
Felwort, Marsh 111
Flax, Alpine 107
Fleabane, Alpine 39
Fleabane, One-flowered 90
Fleawort, Field 39
Forget-me-not, Alpine 112
Foxglove, Fairy 113
Foxglove, Large Yellow 170
Foxglove, Small Yellow 171
Fragrant Orchid, Short-spurred 50
Fritillary, Alpine 42
Fritillaria tubiformis 42

Gagea liotardii 163
Galium anisophyllon 57
– *megalospermum* 57
Garland Flower 12
Gentian, Alpine 109
Gentian, Bavarian 109
Gentian, Bladder 110
Gentian, Cross 101
Gentian, Dwarf 102
Gentian, Field 14
Gentian, Fringed 101
Gentian, Great Yellow 149
Gentian, Hungarian 33
Gentian, Purple 33
Gentian, Pygmy 120
Gentian, Short-leaved 110
Gentian, Slender 101
Gentian, Snow 109
Gentian, Spotted 149
Gentian, Spring 110
Gentian, Stemless 108
Gentian, Styrian 81

Gentian, Trumpet 108
Gentian, Willow 108
Gentiana acaulis 108
– *alpina* 109
– *asclepiadea* 108
– *bavarica* 109
– *brachyphylla* 110
– *clusii* 108
– *cruciata* 101
– *frigida* 81
– *lutea* 149
– *nivalis* 109
– *pannonica* 33
– *prostrata* 120
– *punctata* 149
– *purpurea* 33
– *utriculosa* 110
– *verna* 110
Gentianella campestris 14
– *ciliata* 101
– *nana* 102
– *tenella* 101
Geranium phaeum 26
– *rivulare* 76
– *sanguineum* 26
– *sylvaticum* 26
Germander, Mountain 98
Germander, Wall 45
Globeflower 148
Globularia, Bald-stemmed 129
Globularia cordifolia 129
– *nudicaulis* 129
Globularia, Matted 129
Goldenrod 156
Good King Henry 176
Goodyera repens 99
Grass-of-Parnassus 75
Groundsel, Grey Alpine 153
Gymnadenia conopsea 49
– *odoratissima* 50
– *rhellicani* 50
– *rubra* 50
Gypsophila, Alpine 62
Gypsophila repens 62

Hare's-ear, Three-veined 146
Harebell, Broad-leaved 119
Hawk's-beard, Alpine 160
Hawk's-beard, Golden 40
Hawk's-beard, Kerner's 160
Hawk's-beard, Large-headed 159
Hawk's-beard, Pygmy 161
Hawk's-beard, Pyrenean 159
Hawk's-beard, Triglav 160
Hawkbit, Mountain 158

Hawkbit, Rough 158
Hawkbit, Swiss 158
Hawkweed, Alpine 162
Hawkweed, Chicory-leaved 161
Hawkweed, Mouse-ear 161
Hawkweed, Orange 40
Hawkweed, Silver 162
Heath, Spring 11
Heather 31
Hedysarum hedysaroides 44
Helianthemum alpestre 140
– *nummularium* 139
Helleborine, Broad-leaved 179
Helleborine, Dark-red 49
Helleborine, Marsh 99
Helleborine, White False 92
Helleborus niger 85
Hepatica 120
Hepatica nobilis 120
Heracleum austriacum 78
– *sphondylium* 78
Herb Paris 179
Herminium monorchis 180
Hieracium alpinum 162
– *aurantiacum* 40
– *intybaceum* 161
– *pilosella* 161
– *villosum* 162
Hippocrepis comosa 168
Hogweed 78
Hogweed Austrian, 78
Homogyne alpina 35
Honeywort, Smooth 147
Horminum pyrenaicum 128
Houseleek, Alpine 32
Houseleek, Cobweb 32
Houseleek, Hen-and-chickens 148
Houseleek, Mountain 32
Houseleek, Wulfen's 149
Hugueninia tanacetifolia 134
Hypericum maculatum 139
Hypochoeris uniflora 157

Iberis saxatilis 55

Jacobaea abrotanifolia 154
– *incana* 153
Jacob's Ladder 111

Kernera, Rock 55
Kernera saxatilis 55
Kidney Vetch, Alpine 167
King of the Alps 112
Knapweed, Alpine Greater 36

Knapweed, Plumed 37
Knapweed, Singleflower 36
Knapweed, Squarrose 121
Knapweed, Wig 37
Knotgrass, Alpine 65
Knautia dipsacifolia 131
– *longifolia* 131

Lady's-mantle, Alpine 176
Lady's-mantle, Common 177
Lady's-slipper 174
Lady's-tresses, Creeping 99
Lamium album 97
– *galeobdolon* 169
Larkspur, Alpine 106
Laserpitium halleri 77
– *latifolium* 77
– *siler* 78
Lathyrus laevigatus 168
– *pratensis* 168
Leek, Alpine 94
Lentil, Mountain 167
Leontodon hispidus 158
– *montanus* 158
Leontopodium alpinum 87
Leopard's-bane, Glacier 153
Leopard's-bane, Large-flowered 152
Leopard's-bane, Tufted 152
Lettuce, Purple 39
Leucanthemopsis alpina 90
Leucanthemum adustum 90
Ligusticum mutellinoides 27
– *mutellina* 27
Lilium bulbiferum 41
– *martagon* 42
Lily, Dog's-tooth 41
Lily, Fire 41
Lily, Martagon 42
Lily, Snowdon 94
Lily, St Bernard's 93
Lily, St Bruno's 93
Lily-of-the-valley 92
Linaria alpina 129
Linnaea borealis 82
Linum alpinum 107
Listera cordata 180
– *ovata* 180
Lloydia serotina 94
Loiseleuria procumbens 19
Lomatogonium carinthiacum 111
Lomatogonium, Carinthian 111
Lotus alpinus 166
– *corniculatus* 166
Lousewort, Beaked 47

Lousewort, Crested 173
Lousewort, Crimson-tipped 173
Lousewort, Flesh-pink 47
Lousewort, Kerner's 46
Lousewort, Leafy 173
Lousewort, Pink 47
Lousewort, Truncate 46
Lousewort, Tuberous 174
Lousewort, Whorled 48
Lovage, Alpine 27
Lovage, Unbranched 27
Lungwort, South-alpine 113

Marguerite, Mountain 91
Marsh-marigold 137
Marsh-orchid, Broad-leaved 48
Marsh-orchid, Early 49
Masterwort 79
Masterwort, Great 76
Masterwort, Lesser 77
Meadow-rue, Alpine 10
Meadow-rue, Greater 10
Meadow-rue, Lesser 132
Melampyrum pratense 171
Menyanthes tifoliata 81
Meum athamanticum 80
Milk-vetch, Alpine 96
Milk-vetch, Mountain 125
Milk-vetch, Southern 96
Milk-vetch, Yellow Alpine 97
Milkwort, Alpine 126
Milkwort, Bitter 126
Milkwort, Shrubby 169
Minuartia sedoides 178
– *verna* 62
Moehringia ciliata 62
Moneses uniflora 66
Monk's Rhubarb 30
Monk's-hood 123
Monk's-hood, Variegated 123
Moon-daisy, Alpine 90
Mountain Everlasting 35
Mountain Sorrel 10
Mountain Tea, Hyssop-leaved 170
Mouse-ear, Alpine 60
Mouse-ear, Broad-leaved 61
Mouse-ear, Field 60
Mouse-ear, Glacier 61
Mouse-ear, Starwort 61
Mullein, White 82
Mustard, Buckler 133
Myosotis alpestris 112

Narcissus poeticus ssp. *radiiflorus* 94
– *pseudonarcissus* 163
Neotinea ustulata 51

Onion, Narcissus 40
Onobrychis montana 44
Ononis cristata 43
Orchid, Burnt 51
Orchid, Black Vanilla 50
Orchid, Early-purple 51
Orchid, Elder-flowered 175
Orchid, False Musk 181
Orchid, Fly 181
Orchid, Fragrant 49
Orchid, Frog 181
Orchid, Musk 180
Orchid, Pale-flowered 175
Orchid, Red Vanilla 50
Orchid, Round-headed 51
Orchid, Small White 175
Orchis mascula 51
– *pallens* 175
Orobanche flava 174
Orphrys insectifera 181
Orthilia secunda 66
Ox-eye, Splendid Heartleaf 157
Ox-eye, Yellow 157
Oxlip 141
Oxyria digyna 10
Oxytropis campestris 167
– *halleri* 126
– *jacquinii* 125
Oxytropis, Purple 126
Oxytropis, Yellow 167

Paederota, Bluish 105
Paederota bonarota 105
– *lutea* 136
Paeonia officinalis 30
Pansy, Alpine 124
Pansy, Long-spurred 123
Pansy, Mount Cenis 124
Pansy, Mountain 165
Pansy, Wild 165
Papaver alpinum ssp. *rhaeticum* 132
– *alpinum* ssp. *sendtneri* 52
Paradisea liliastrum 93
Parageum montanum 150
– *reptans* 150
Paris quadrifolia 179
Parnassia palustris 75
Pasqueflower, Alpine 85
Pasqueflower, Mountain 29
Pasqueflower, Spring 85

Index

Pasqueflower, Yellow Alpine 148
Pea, Yellow 168
Pearlwort, Alpine 63
Pedicularis comosa 173
– *foliosa* 173
– *kerneri* 46
– *oederi* 173
– *recutita* 46
– *rosea* 47
– *rostratocapitata* 47
– *rostratospicata* 47
– *tuberosa* 174
– *verticillata* 48
Penny-cress, Alpine 56
Penny-cress, Round-leaved 11
Peony, Common 30
Petasites paradoxus 34
Petrocallis pyrenaica 11
Peucedanum ostruthium 79
Pheasant's-eye 94
Physoplexis comosa 28
Phyteuma betonicifolium 116
– *globulariifolium* 114
– *hemisphaericum* 114
– *orbiculare* 114
– *ovatum* 115
– *scheuchzeri* 115
– *sieberi* 115
Pimpinella major 27
Pinguicula alpina 98
– *vulgaris* 130
Pink, Alpine 15
Pink, Glacier 15
Pink, Large 16
Pink, Wood 16
Plantain, Alpine 177
Plantain, Dark 177
Plantain, Greater 178
Plantain, Ribwort 178
Plantago alpina 177
– *atrata* 177
– *lanceolata* 178
– *major* 178
Platanthera bifolia 99
Polemonium caeruleum 111
Polygala alpestris 126
– *amara* 126
– *chamaebuxus* 169
Polygonatum verticillatum 92
Polygonum alpinum 65
Poppy, Alpine 52
Poppy, Yellow Alpine 132
Potentilla aurea 145
– *caulescens* 75
– *clusiana* 75

– *crantzii* 145
– *erecta* 136
– *frigida* 145
– *grandiflora* 146
– *nitida* 25
Prenanthes purpurea 39
Primrose, Bird's-eye 21
Primrose, Entire-leaved 22
Primrose, Hairy 22
Primrose, Haller's 22
Primrose, Least 23
Primrose, Piedmont 23
Primrose, Splendid 23
Primrose, Sticky 107
Primula auricula 140
– *elatior* 141
– *farinosa* 21
– *glutinosa* 107
– *halleri* 22
– *hirsuta* 22
– *integrifolia* 22
– *minima* 23
– *pedemontana* 23
– *spectabilis* 23
– *veris* 141
Pritzelago alpina 56
Prunella grandiflora 128
Pseudolysimachion spicatum 105
Pseudorchis albida 175
Pulmonaria australis 113
Pulsatilla alpina ssp. *alpina* 85
– *alpina* ssp. *apiifolia* 148
– *montana* 29
– *vernalis* 85
Pyrola rotundifolia 67

Ragwort, Alpine 154
Ragwort, Chamois 153
Ragwort, Gaudin's 155
Ragwort, Rock 154
Ragwort, Southernwood 154
Rampion, Blue-spiked 116
Rampion, Dark 115
Rampion, Dolomite 115
Rampion, Globe-headed 114
Rampion, Horned 115
Rampion, Round-headed 114
Rampion, Round-leaved 114
Ranunculus aconitifolius 58
– *acris* 138
– *alpestris* 59
– *flammula* 137
– *glacialis* 58
– *hybridus* 137

– *kuepferi* 59
– *lanuginosus* 138
– *montanus* 138
– *parnassifolius* 59
– *platanifolius* 58
– *repens* 139
Restharrow, Mount Cenis 43
Rhinanthus alectorolophus 172
– *glacialis* 172
Rhodiola rosea 135
Rhododendron ferrugineum 19
– *hirsutum* 19
Rhodothamnus chamaecistus 20
Rock-cress, Alpine 52
Rock-cress, Bluish 100
Rock-cress, Dwarf Alpine 52
Rock-cress, Shiny 53
Rocket, Nasturtium-leaved Hairy 134
Rocket, Tansy-leaved 134
Rock-jasmine, Alpine 20
Rock-jasmine, Blunt-leaved 69
Rock-jasmine, Dolomite 69
Rock-jasmine, Flesh-pink 20
Rock-jasmine, Hairy 68
Rock-jasmine, Milkwhite 68
Rock-jasmine, Sweetflower 68
Rock-jasmine, Swiss 69
Rock-jasmine, Yellow 140
Rock-rose, Alpine 140
Rock-rose, Common 139
Roseroot 135
Rumex acetosella 30
– *alpestris* 29
– *alpinus* 30
– *nivalis* 29
– *scutatus* 176

Saffron, Alpine 42
Sage, Sticky 169
Sagina saginoides 63
Sainfoin, Alpine 44
Sainfoin, Mountain 44
Salvia glutinosa 169
Sandwort, Creeping (Ciliate) 62
Sandwort, Fringed 60
Sandwort, Spring 62
Sandwort, Two-flowered 64
Sanguisorba officinalis 12
Saponaria ocymoides 16
– *pumila* 17

Saussurea alpina 121
– *discolor* 122
– *pygmea* 122
– *rhapontica* 37
Saussurea, Dwarf 122
Saussurea, Heart-leaved 122
Saw-wort, Alpine 121
Saw-wort, Alpine 37
Saxifraga aizoides 143
– *androsacea* 73
– *aphylla* 144
– *arachnoidea* 73
– *aspera* 72
– *biflora* 24
– *bryoides* 143
– *burseriana* 72
– *caesia* 72
– *cotyledon* 70
– *crustata* 71
– *cuneifolia* 74
– *hostii* 71
– *moschata* 143
– *muscoides* 74
– *mutata* 25
– *oppositifolia* 25
– *paniculata* 71
– *rotundifolia* 73
– *sedoides* 144
– *seguieri* 144
– *stellaris* 74
Saxifrage, Blue-grey 72
Saxifrage, Burser's 72
Saxifrage, Cobweb 73
Saxifrage, Eastern 144
Saxifrage, Encrusted 71
Saxifrage, Flat-leaved 74
Saxifrage, Host's 71
Saxifrage, Leafless 144
Saxifrage, Live-long 71
Saxifrage, Moss 143
Saxifrage, Musky 143
Saxifrage, Orange 25
Saxifrage, Purple 25
Saxifrage, Pyramidal 70
Saxifrage, Rough 72
Saxifrage, Round-leaved 73
Saxifrage, Scree 73
Saxifrage, Seguier's 144
Saxifrage, Shield-leaved 74
Saxifrage, Starry 74
Saxifrage, Two-flowered 24
Saxifrage, Yellow Mountain 143
Scabiosa lucida 131
Scabious, Devils-bit 105
Scabious, Long-leaved 131

Scabious, Shining 131
Scabious, Wood 131
Scorpion Vetch, Small 166
Scorzoneroides helvetica 158
Scutellaria alpina 128
Sedum acre 142
– *album* 70
– *alpestre* 141
– *anacampseros* 24
– *annuum* 142
– *atratum* 24
– *dasyphyllum* 70
– *montanum* 142
Self-heal, Large 128
Sempervivum arachnoideum 32
– *globiferum* 148
– *montanum* 32
– *tectorum* ssp. *alpinum* 32
– *wulfenii* 149
Senecio abrotanifolius 154
– *alpinus* 154
– *doronicum* 153
– *rupestris* 154
Sermountain, Broad-leaved 77
Sermountain, Haller's 77
Sermountain, Narrow-leaved 78
Seseli libanotis 80
Sibbaldia 146
Sibbaldia procumbens 146
Sideritis hyssopifolia 170
Silene acaulis 17
– *dioica* 17
– *nutans* 64
– *pusilla* 63
– *rupestris* 63
– *suecica* 18
– *vulgaris* 64
Skullcap, Alpine 128
Snowbell, Alpine 31
Snowbell, Dwarf 31
Snowbell, Least 86
Soapwort, Dwarf 17
Soapwort, Rock 16
Soldanella alpina 31
– *minima* 86
– *pusilla* 31
Solidago virgaurea ssp. *alpestris* 156
Solomon's-seal, Whorled 92
Sorrel, French 176
Sorrel, Sheep's 30
Sow-thistle, Alpine 122
Spearwort, Lesser 137

Speedwell, Alpine 103
Speedwell, Compact Blue 104
Speedwell, Germander 102
Speedwell, Heath 104
Speedwell, Leafless-stemmed 103
Speedwell, Rock 102
Speedwell, Spiked 105
Speedwell, Sub-shrub 14
Speedwell, Thyme-leaved 104
Speedwell, Violet 103
Spignel 80
Spotted-orchid, Heath 48
Spurge, Cypress 132
Stachys alopecuros 170
– *alpina* 45
– *pradica* 45
St John's-wort, Imperforate 139
Star-of-Bethlehem, Liotard's 163
Stonecrop, Alpine 141
Stonecrop, Annual 142
Stonecrop, Biting 142
Stonecrop, Dark 24
Stonecrop, Love-restoring 24
Stonecrop, Mountain 142
Stonecrop, Thick-leaved 70
Stonecrop, White 70
Streptopus amplexifolius 93
Succisa pratensis 105
Sweet William 15
Swertia perennis 111

Tanacetum corymbosum 90
Tansy, Corymbose 90
Taraxacum sect. Alpina 159
Telekia speciosissima 157
Tephroseris integrifolia ssp. *capitata* 39
– *longifolia* ssp. *gaudinii* 155
Teucrium chamaedrys 45
– *montanum* 98
Thalictrum alpinum 10
– *aquilegiifolium* 10
– *minus* 132
Thesium alpinum 56
– *pyrenaicum* 76
Thistle, Alpine 35
Thistle, Dwarf 38
Thistle, Great Marsh 36
Thistle, Melancholy 38
Thistle, Spiniest 151
Thistle, Stemless Carline 86
Thistle, Woolly 38
Thistle, Yellow Melancholy 150

Index

Thlaspi caerulescens 56
– *cepaeifolium* 11
Three-leaved Valerian 83
Thrift, Alpine 18
Thyme, Hairy 46
Thymus praecox ssp. *polytrichus* 46
Toadflax, Alpine 129
Tofieldia calyculata 162
– *pusilla* 91
Tormentil 136
Tozzia alpina 172
Tozzia, Alpine 172
Tragacanth, Mountain 97
Traunsteinera globosa 51
Treacle-mustard, Swiss 135
Trifolium alpinum 43
– *badium* 165
– *montanum* 95
– *palescens* 95
– *pratense* ssp. *nivale* 95
– *rubens* 43
– *thalii* 96
Trollius europaeus 148
Tulipa sylvestris ssp. *australis* 163
Tulip, Wild 163
Tussilago farfara 152
Twayblade, Common 180
Twayblade, Lesser 180
Twinflower 82
Twisted-stalk 93

Vaccinium myrtillus 179
– *uliginosum* ssp. *pubescens* 67
– *vitis-idaea* 67
Valerian, Alpine 147
Valerian, Dwarf 28
Valerian, Mountain 28

Valerian, Rock 83
Valerian, Scree 83
Valeriana celtica 147
– *montana* 28
– *saliunca* 83
– *saxatilis* 83
– *supina* 28
– *tripteris* 83
Veratrum album 92
Verbascum lychitis 82
Veronica allionii 104
– *alpina* 103
– *aphylla* 103
– *bellidioides* 103
– *chamaedrys* 102
– *fruticans* 102
– *fruticulosa* 14
– *officinalis* 104
– *serpyllifolia* 104
Veronica, Yellow 136
Vetch, Horseshoe 168
Vetch, Tufted 125
Vetchling, Meadow 168
Vicia cracca 125
Vincetoxicum, Common 82
Vincetoxicum hirundinacea 82
Viola alpina 124
– *biflora* 164
– *calcarata* 123
– *cenisia* 124
– *lutea* 165
– *palustris* 125
– *pinnata* 124
– *tricolor* 165
Violet, Finger-leaved 124
Violet, Marsh 125
Violet, Yellow Wood 164

Whitlowgrass, Carinthian 54
Whitlowgrass, Hoppe's 135
Whitlowgrass, Pyrenean 11
Whitlowgrass, Woolly 55
Whitlowgrass, Yellow 134
Willowherb, Alpine 14
Willowherb, Chickweed 13
Willowherb, Fleischer's 13
Willowherb, Rosebay 13
Wintergreen, One-flowered 66
Wintergreen, Round-leaved 67
Wintergreen, Serrated 66
Wolfsbane, Northern 164
Wormwood 155
Wormwood, Alpine 156
Wormwood, Glacier 155
Wormwood, Spiked 156
Woundwort, Alpine 45
Wulfenia carinthiaca 113
Wulfenia, Carinthian 113

Yarrow 87
Yarrow, Black 88
Yarrow, Broad-leaved 88
Yarrow, Dwarf Alpine 89
Yarrow, Musk 89
Yarrow, Silvery 88
Yellow-rattle, Awned 172
Yellow-rattle, Greater 172

The Author, Ansgar Hoppe

Plants have always been important to Ansgar Hoppe. After working in forestry, Ansgar Hoppe studied biology at the University of Osnabrück and gained his PhD in botany and vegetation science at the Institute of Geobotany at the University of Hanover. He learned about the plants of the Alps and other European mountains through numerous excursions. Since then he has been a scientific associate at the University and the Niedersächsischer Heimatbund (NHB) [Heritage Association of Lower Saxony] in Hanover. He is involved with many projects researching conservation of the cultural landscape, as well as working freelance as a botanical expert and author.

Photo credits

Front cover: A large-flowered gentian (Christa Eder/fotolia.com)

Back cover:
Dark Stonecrop (left: Ansgar Hoppe)
Yellow Mountain Saxifrage (right: Ansgar Hoppe)
Hairy Rock-jasmine (middle: Manuel Werner)

516 photographs: 1 by Harald Berger (65 m), 1 by Filip Dominec (98 b), 2 by Gartenschatz (102 m, 120 t), 23 by Edmund Garnweidner, 1 by Franz Hadacek (102 t), 1 by Bernd Haynold (123 m), 146 by Michael Hassler, 32 by Frank Hecker, 3 by Ansgar Hoppe (24 m, 116 b, 143 t), 1 by Ernst Horak (83 t), 1 by Hansjörg Küster (39 b, 189), 1 by Hans E. Laux (8/9), 1 by Kjetil Lenis (165 t), 2 by Frantisek Pleva (67 t, 76 m), 1 by Hermann Schachner (65 b), 3 by Johann Schneidert (24 b, 122 t, 145 b), 4 by Herbert Wagner (53 m, 69 t, 118 t, 122 m) and 1 by Manuel Werner (68 b).
All other photographs are by Peter Mertz, naturalist.
47 photographs: 1 drawing and 1 map at the end of the book by Wolfgang Lang.

© 2012, Franckh-Kosmos Verlags-GmbH & Co. KG, Stuttgart.

Published by Pelagic Publishing, PO Box 725, Exeter, EX1 9QU, UK

www.pelagicpublishing.com

ISBN: 978-1-907807-40-4

Technical terms illustrated

Flower

radially symmetrical (actinomorphic)

tip (of sepal)

petals free   petals fused

corolla fused, bell-shaped

bilaterally symmetrical (zygomorphic) e.g.

Legume flower

standard
wing
keel

Labiate flower

Upper lip
Lower lip

composite flower

Disc- (tube-) floret   Ray-floret

Orchid flower

Spur
Lip
Bract

Inner disc-florets

(involucral) bract

Outer ray-florets

Flowerhead

Stigma
Style
Petal
Anther (of Star
Ovary
Sepal

Spike   Raceme   Panicle   Umbel with bracts   Flowerhead

Compound umbel with bracts and bracteoles

Corymb

Umbel-like panicle

Technical terms illustrated

Leaf shapes

Linear

Lanceolate

Ovate

Obovate

Kidney-shaped

Heart-shaped

Trifoliate

Palmate

Basal rosette

Imparipinnate

Paripinnate with terminal tendrils

Bipinnate

Leaf with stipule

Leaf margins

Entire   Serrate   Crenate (round-toothed)   Toothed

Leaf arrangement

Crossed opposite (decussate opposite)

Whorled   Alternate

About Pelagic Publishing
We publish books for scientists, conservationists, ecologists, wildlife enthusiasts – anyone with a passion for understanding and exploring the natural world.

www.pelagicpublishing.com